28 Advances in Polymer Science

Fortschritte der Hochpolymeren-Forschung

Edited by H.-J. CANTOW, Freiburg i. Br. · G. DALL'ASTA, Colleferro
K. DUŠEK, Prague · J. D. FERRY, Madison · H. FUJITA, Osaka
M. GORDON, Colchester · W. KERN, Mainz · G. NATTA, Milano
S. OKAMURA, Kyoto · C. G. OVERBERGER, Ann Arbor · T. SAEGUSA, Kyoto
G. V. SCHULZ, Mainz · W. P. SLICHTER, Murray Hill · J. K. STILLE, Fort Collins

With 39 Figures

Springer-Verlag
Berlin Heidelberg GmbH 1978

Editors

ISBN 978-3-662-15452-6   ISBN 978-3-540-35798-8 (eBook)
DOI 10.1007/978-3-540-35798-8

Library of Congress Catalog Card Number 61–642

Contents

Random and Block Copolymers by Ring-Opening Polymerization

Yuya Yamashita

Department of Synthetic Chemistry, Faculty of Engineering, Nagoya University, Nagoya 464, Japan

Table of Contents

1. Introduction

Extensive literature on the polymerization of cyclic monomers is available but is to a large extent descriptive of the modes of the polymerizations and the properties of the resulting products. It is only in the past twenty years that the individual results have been summarized to a system of ring-opening polymerization. Recent trends of polymer chemistry indicate that copolymer systems become increasingly important when searching for optimum properties. This review is primarily concerned with the syntheses of copolymers by ring-opening polymerization. The structures of random and alternating copolymers obtained by this kind of polymerization provide significant information regarding the nature of the propagation reactions. The syntheses of block and graft copolymers proved the living nature of the propagation reaction and also provided valuable information for the design of tailor-made polymers which possess specific properties.

2. General Features of Ring-Opening Polymerization

Ring-opening polymerization can be characterized by three general features. The first feature is the frequent observation of the polymerization equilibria which result from a decreased polymerizability of cyclic monomers compared with vinyl monomers. The second is the tendency to form living polymers, which is attributable to the stability of the propagating species. The third feature is a tendency toward depolymerization or ring-chain equilibration of heterochain polymers. These features play an important role in determining the structure of random and block copolymers obtained by ring-opening polymerization.

a) Polymerizability of Cyclic Monomers

A considerable body of information is available about the polymerizability of cyclic monomers, part of it is illustrated in Table 1 [1−5]. The thermodynamic driving forces toward the polymerization of a cyclic monomer are angular strains for three- and four-membered rings and steric interactions between hydrogen atoms for monomers with more than five ring atoms. The heats of polymerization of three- and four-membered ring monomers are as large as 20 kcal/mole and these monomers polymerize as readily as vinyl monomers. Steric interaction between quasi-axial hydrogen atoms becomes important for cyclic monomers with more than seven-membered ring atoms; their heats of polymerization decrease to about or below 10 kcal/mole. Because the free energies of polymerization of such monomers are negative, due to a small contribution of the entropy term at ordinary temperature, formation of polymers can be expected. The probability of ring closure decreases for large membered rings and the entropy increase becomes the driving force for the polymerization of cyclic monomers with more than twelve-membered ring atoms [5, 6, 73].

Table 1. Polymerizability (+ or −) of cyclic monomers

Class of Monomer	3	4	5	6	7	8 and higher
Ethers	+	+	+	−	+	
Formals			+	−	+	+
Lactones		+	−	+	+	+
Carbonates			−	+	+	
Anhydrides			−	−	+	+
Imines	+	+	−	−	+	
Lactams		+	+	+	+	+
Urethanes			−	+	+	
Ureas			+	−	+	
Imides			−	−	+	
Sulfides	+	+	−	−		
Disulfides		+	+	+	+	+
Sultones		+	+	−		

Cyclic monomers with five- and six-membered ring atoms exist in strainless puckered conformations; their heats of polymerization are either negative or have small positive values due to the repulsion of eclipsed hydrogen atoms. Because the enthalpy and entropy contributions are comparable, the free energies of polymerization are either positive or may become positive at high temperatures.

b) Kinetics of Ring-Opening Polymerization

The propagating species involved in ring-opening polymerizations are more stable and hence less reactive than those in vinyl polymerization. Typically they are alkoxide anions or carboxylate anions in anionic polymerization systems and oxonium ions, sulfonium ions or immonium ions in cationic polymerization systems. The mechanism of ring-opening propagation reactions is a substitution reaction of the S_N2 type rather than the addition reaction in vinyl polymerization. It has been shown in radical polymerization systems that the extent of side reactions, such as chain transfer and chain termination, decreases relative to the propagation reactions as the stability of the propagating species increases.

Although the reactivities of both the propagating species and the monomers affect the rate of polymerization, the values shown in Table 2 suggest that the former factor is more important[24−35]. Such a trend is familiar in radical polymerization. In this kind of polymerization, non-conjugated monomers, which tend to give a less stabilized propagating radical, polymerize at faster rates than do conjugated monomers, which give resonance-stabilized propagating radicals[36]. In ionic polymerization systems, the existence of various states of the propagating species is generally accepted and the rate constants for the propagation reactions differ widely from dissociated ion pairs to associated ion pairs. The existence of a highly reactive free ion or solvated ion pairs was observed in the polymerization of propylene sulfide[37], dimethyl thietane[38], ethylene oxide[39], heamethylcyclotrisiloxane[40],

Table 2. Rate constants of propagation for ionic polymerization

Propagation Reaction	Solvent	Temperature °C	kp (1/mole sec)	Ref.
St^+ + St	CH_2Cl_2	−80	2200	24)
DOL^+ + DOL	CH_2Cl_2	0	100	25)
OX^+ + OX	CH_2Cl_2	0	0.14	26)
THF^+ + THF	CH_2Cl_2	0	0.005	27)
DMT^+ + DMT	CH_2Cl_2	20	0.0065	28)
$MOXZ^+$ + MOXZ	CH_3CN	40	0.0001	29)
St^- + St	THF	25	550	30)
RS^- + PS	THF	−30	0.04	31)
RO^- + EO	THF	50	0.02	32)
$RCOO^-$ + PVL	CH_3CN	35	0.12	33)
RNH_2 + Bz-L-G-NCA	DOX	25	0.03	34)
ROH + GAS	$C_6H_5CH_2OH$	90	0.0004	35)

DOL: 1.3-Dioxolane, OX: oxetane, THF: tetrahydrofuran.
DMT: 3.3-Dimethylthietane, MOXZ: 2-methyl-2-oxazoline.
PS: Propylene sulfide, EO: ethylene oxide.
PVL: Pivalolactone, Bz-L-G-NCA: γ-benzyl-L-glutamate NCA.
GAS: Glycolic acid anhydrosulfite.

tetrahydrofuran[41] and 2-phenyloxazoline[42]. However, the observed rate of the free ion is sometimes only ten times greater than that of the solvated ion pair, compared with the factor of 10^3 in the anionic polymerization of hydrocarbon monomers[30]. In some cases the solvated ion pair was found to be more reactive than the free ion. This was attributed to the fact that the mechanism of propagation by the ion pair differs from that of the free ion[43]. Together with ionic species, covalent species, such as ester[44, 45] or iodide[46] in cationic systems and amine[16] or alcohol[35] in anionic systems, sometimes become active species for the propagation reaction. The rate constant for the propagation depends on the concentration of propagating species, the relative amounts of which are determined by the polarity of the solvent and by the nature of the catalyst. In spite of these factors, the general trend observed in Table 2 agrees with the order of reactivity of the propagating species. In cationic polymerization this order is the electrophilicity of the propagating species, $C^+ > O^+ > S^+ > N^+$, and in anionic polymerizations, it is the nucleophilicity of the propagating species, $C^- > S^- > O^- > COO^- > NH_2 > OH$.

Since the discovery of a living polymer in the anionic polymerization of styrene[30], extensive studies have been carried out to find new living systems. Although the propagating carbocation in the cationic polymerization of styrene is so unstable that elimination of β-hydrogen atoms readily occurs, the propagating species in the cationic polymerization of tetrahydrofuran is a stable oxonium ion, so that the polymerization of tetrahydrofuran proceeds as a living system with practically no chain transfer and no termination if the catalyst is properly selected[47]. It is necessary to confirm that the concentration of the living end does not decrease during the polymerization in order to conclude the absence of chain termination[48].

It is also necessary, in order to conclude the absence of chain transfer[49], to observe that the molecular weight of the polymer produced increases linearly with conversion.

For the cationic polymerization of other cyclic oxygen compounds the chain transfer or termination reaction was observed with 3,3-bischloromethyloxetane[50], 1,3-dioxolane[51] and ε-caprolactone[52], although an initial increase in molecular weight with conversion was sometimes noticed. In the case of oxiranes[53], thiiranes[54] and aziridines[55], chain transfer to polymer is important. Cationic polymerization of unsubstituted oxazolines[56], substituted thietanes[57] and azetidines[57] presented the possibility of obtaining new living polymers by suppressing chain transfer and termination reactions.

More examples of the living polymer formation can be found in anionic polymerization systems. The polymerization of propylene sulfide[58] is a typical example. The polymerization of propylene oxide is accompanied by monomer transfer reactions[59], whereas ethylene oxide forms a living polymer[60]. The absence of termination and chain transfer reactions was confirmed in the anionic polymerization of pivalolactone[33] and tetramethylglycolide[61], but considerable amounts of cyclic oligomer were formed in the case of ε-caprolactone[62]. Also, α-pyrrolidone[63], N-carboxy-α-amino acid anhydrides[64], hexamethylcyclotrisiloxane[65] and chloral[66] are known to give living polymers. Further studies seem necessary to confirm the formation of living polymers in the anionic polymerization of isocyanates and dimethylketene. Living polymers are also formed in complex catalyst systems. The recent discovery that the aluminum alkoxide catalyzed polymerization of ε-caprolactone or β-propiolactone yielded a living polymer is quite interesting[67-69].

Monodisperse polymers can be obtained in living systems if the initiation is fast relative to the propagation. This was the case with tetrahydrofuran[70, 71]. Limited experiments at preparing low molecular weight monodisperse polymers were reported for ethylene oxide[60], hexamethylcyclotrisiloxane[65] and sarcosine NCA[72].

c) Depolymerization and Ring-Chain Equilibrium

Heterochain polymers produced by ring-opening polymerization contain the heteroatoms in the main chain as well as in the monomer and the polymer chain competes with the monomer for the reaction with the propagating species. This competition leads to polymer transfer and back-biting reactions during the polymerization. Heterochain polymers are also susceptible to depolymerization by the ionic active species which are easily formed during processing.

The existence of cyclic oligomers in equilibrium with condensation polymers was noticed earlier. The distribution of large cyclic oligomers formed by ring-opening polymerization, when analyzed by chromatography, agreed with the expected value from the Jacobson-Stockmayer theory[73, 74], which is based on the assumption that the equilibrium cyclization constant depends on the probability of coincidence of chain ends determined by the conformation of the polymer chain. Examples of agreement are polysiloxanes[75], polydioxolanes[76], polyamides[77] and polyesters[78]. The selective formation of small cyclic oligomers was reported on the polymerization of epoxides[79, 80], oxetanes[81], cyclic formals[82, 83], lactones[60, 84], episulfides[85] and aziridines[86]. Several mechanisms were postulated to explain the formation of such cyclic oligomers[87−89].

The rate of the depolymerization of heterochain polymers depends on the structure of the polymers and the nature of the catalysts. The depolymerization results in a redistribution of polymer chains by chain cleavage, besides the formation of monomer and oligomers. Equilibration of polysiloxane oligomers by acidic or basic catalysts is used commercially in the silicone industry. Redistribution of polyacetals[90], polyethers[91, 92] and polyesters[93] occurs slowly at room temperature and changes the molecular weight distribution. Unzipping from the living chain end occurs frequently and it is sometimes necessary to kill the living end when the polymer is isolated from the reaction mixture. Otherwise, the polymer obtained may become unstable at high temperature and depolymerization may occur giving an equilibrium mixture of monomers and oligomers. Polyacetal is thermally unstable, but stabilization can be achieved by end capping or by copolymerization. Depolymerization also occurs in the case of polysters for which acetylation of the chain end is effective in enhancing their thermal stability[94, 95]. The carboxylate end groups of polyesters become more stable if the counter cation is changed from alkali metal to quaternary ammonium salts[96].

3. Random and Alternating Copolymers

a) Mechanism and Kinetics of Copolymerization

Extensive work on radical copolymerization has shown that the composition in a binary copolymer, consisting of monomers M_1 and M_2, is determined by four rate constants k_{ij} for a propagating chain ending with M_i adding to monomer M_j.

$$
\begin{aligned}
M_1^* + M_1 &\xrightarrow{k_{11}} M_1^* \\
M_1^* + M_2 &\xrightarrow{k_{12}} M_2^* \\
M_2^* + M_1 &\xrightarrow{k_{21}} M_1^* \\
M_2^* + M_2 &\xrightarrow{K_{22}} M_2^*
\end{aligned}
$$

The Mayo-Lewis equation expressing the copolymer composition can be derived from these four elementary reactions. It reads

$$
\frac{d\,(M_1)}{d\,(M_2)} = \frac{(M_1)\,[r_1\,(M_1) + (M_2)]}{(M_2)\,[(M_1) + r_2\,(M_2)]}
$$

where the parameters $r_1 = k_{11}/k_{12}$ and $r_2 = k_{22}/k_{21}$ are called the monomer reactivity ratios.

However, ionic copolymerizations are much more selective than radical copolymerizations and the number of copolymer pairs which undergo ionic copolymerization is relatively limited. Cross-propagation rarely occurs between monomer pairs with different reactivities. The most important factor for the formation of a random copolymer is likely to be the similarity of the propagating species. Selective polymerization of a more reactive monomer with a more stable propagating species is a general phenomenon. But the formation of homopolymer mixtures is sometimes observed for monomer pairs with different propagating species. Typical examples are the reactions N-vinylcarbazole with oxetane[97], α-methylstyrene with β-propiolactone[98] and tetrahydrofuran with 1,3-dioxolane[99]. O'Driscoll[100] derived a non-stationary copolymerization equation based on the assumption that the copolymer composition is determind by the ratio of the initiation rates and the ratio of the propagation rates with cross-over reactions being negligible. In the equations which follow A and P represent the initiator and the polymer chain, respectively.

Initiation $\quad A^* \; + M_1 \xrightarrow{\;k_1\;} AM_1^*$

$\qquad\qquad\quad A^* \; + M_2 \xrightarrow{\;k_2\;} AM_2^*$

Propagation $A_1^* \; + M_1 \xrightarrow{\;k_{11}\;} PM_1^*$

$\qquad\qquad AM_2^* + M_2 \xrightarrow{\;k_{22}\;} PM_2^*$

$$\frac{d\,(M_1^*)}{dt} = k_1\,(A^*)\,(M_1), \qquad\qquad \frac{d\,(M_2^*)}{dt} = k_2\,(A^*)\,(M_2)$$

Thus $\qquad \dfrac{d\,(M_1^*)}{d\,(M_2^*)} = \dfrac{k_1\,(M_1)}{k_2\,(M_2)} = \dfrac{(M_1^*)}{(M_2^*)} \qquad$ so that

$$-\frac{d\,(M_1)}{dt} = k_{11}\,(M_1^*)\,(M_1), \qquad\qquad -\frac{d\,(M_2)}{dt} = k_{22}\,(M_2^*)\,(M_2)$$

and by division we obtain

$$\frac{d\,(M_1)}{d\,(M_2)} = \frac{k_{11}\,(M_1^*)\,(M_1)}{k_{22}\,(M_2^*)\,(M_2)} = \frac{k_{11}k_1\,(M_1)^2}{k_{22}k_2\,(M_2)^2}$$

Thus a log-log plot of the apparent copolymer composition ratio against the monomer feed ratio should yield a straight line with slope two. In fact, a slope near two was observed for several comonomer pairs with different polarities.

Thus, confirmation of whether the product obtained in an attempted reaction in a true random copolymer is important to clarify the mechanism of the propagation reaction and to correlate structure and reactivity in ring-opening polymerizations. Considering that apparent copolymers may be formed by reactions other than copolymerization, for example, by ionic grafting or by combination of polymer chains, characterization of cross-sequences appears to be one of the best ways to check the formation of random copolymers.

As an example, an NMR spectrum of a 1,3-dioxolane-β-propiolactone copolymer, obtained by using a boron-fluoride catalyst, is shown in Fig. 1[101]. The 1,3-dioxolane (DOL) homopolymer spectrum contains two singlet peaks of area 1:2 numbered 1 and 5, whereas the spectrum of the β-propiolactone (PL) homopolymer contains two triplet peaks of area 1:1 numbered 2 and 6. Variation of initial feed ratios disclosed that peaks 1,3 and 5 are associated with the DOL units and that

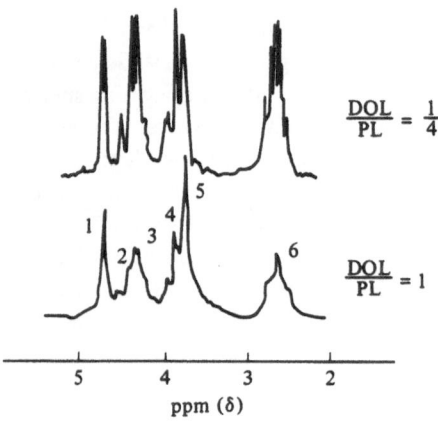

$$\frac{DOL}{PL} = \frac{1}{4}$$

$$\frac{DOL}{PL} = 1$$

ppm (δ)

Fig. 1. NMR spectrum of copolymers of
β-propiolactone (PL) and 1,3-dioxolane (DOL)

peaks 2,4 and 6 are associated with the PL units. Because no peak corresponding
to carbonyl neighbored formal protons $-\underset{\underset{O}{\|}}{C}-O-CH_2-O-$ lower than 5 ppm (δ)
was observed, the structure of the cross-sequence should be
$-OCH_2CH_2COCH_2CH_2OCH_2OCH_2CH_2-$. Peak 3 is reasonably assigned to a carbonyl neighbored oxyethylene unit of DOL and peak 4 to an oxymethylene unit
of PL, which has no PL neighbors. Thus the following assignment to the NMR peaks
is reasonable:

$$-OCH_2CH_2\underset{\underset{O}{\|}}{C}-OCH_2CH_2\underset{\underset{O}{\|}}{C}-OCH_2CH_2OCH_2-OCH_2CH_2OCH_2-OCH_2CH_2\underset{\underset{O}{\|}}{C}-$$

$$2 6 3 3 1_1 5 5 1_2 4 6$$

A cross sequence such as illustrated here may be associated with the following cross
propagation steps,

$$-OCH_2CH_2\underset{\underset{O}{\|}}{C}-^{(+)}O\cdots\underset{\underset{O}{\|}}{C}\cdots O \begin{matrix} CH_2-CH_2 & CH_2-CH_2 \\ & O \\ & CH_2 \end{matrix} \longrightarrow$$

$$-OCH_2CH_2\underset{\underset{O}{\|}}{C}-OCH_2CH_2\underset{\underset{O}{\|}}{C}-^{(+)}O \begin{matrix} CH_2-CH_2 \\ O \\ CH_2 \end{matrix} ---- \begin{matrix} CH_2-CH_2 \\ O-\underset{\underset{O}{\|}}{C} \end{matrix}$$

and, in fact, this was confirmed by the modes of fission of PL[102] and DOL[103].
Another support for the acyl-oxygen fission of PL is the formation of methyl β-
ethoxypropionate in the reaction of PL with triethyloxonium salt followed by
quechning with methanol[104];

$$\text{Et}_3\text{O}^+ + \overset{\overset{\displaystyle \text{CH}_2 - \text{CH}_2}{|\qquad\quad|}}{\text{O} \underset{\underset{\displaystyle \text{O}}{\|}}{\qquad\quad \text{C}}} \quad \xrightarrow{\text{CH}_3\text{OH}} \quad \text{EtOCH}_2\text{CH}_2\text{COOCH}_3$$

Thus, NMR spectroscopy is useful for characterizing the nature of the cross-sequences in random copolymers.

Monomer reactivity ratios are usually determined by treating the data for low conversion copolymers in terms of the Fineman-Ross plot or other graphical methods. A critical reexamination of monomer reactivity ratios in ionic copolymerization systems has shown that the errors involved in these methods are sometimes so large that the monomer reactivity ratios determined are inaccurate and misleading. Improved methods were proposed by Yezrielev[105] and by Tidwell[106−108] but the linear graphic method proposed by Tüdős appears to be simple and reliable[109, 110].

Several important assumptions are involved in the derivation of the Mayo-Lewis equation and care must be taken when it is applied to ionic copolymerization systems. In ring-opening polymerizations, depolymerization and equilibration of the heterochain copolymers may become important in some cases. In such cases, the copolymer composition is no longer determined by the four propagation reactions. In fact, redistribution of copolymer sequences was observed during the copolymerization of trioxane and DOL[111]. Also, chain cleavage of polyether occurred to form apparent copolymers in the copolymerization of epoxides and styrene[112]. Equilibrium copolymerization of cyclic siloxanes has been known for a long time. Accumulation of cyclic oligomers in the early stage of ring-opening polymerizations is frequently noticeable. Monomer reactivity ratios obtained with bulk reaction products is perhaps meaningless without confirming that the products are formed by the true copolymerization reactions[113, 165]. This is important because low conversion products are often mixtures of low molecular weight copolymers and cyclic oligomers which are difficult to separate.

Polymerization equilibria frequently observed in the polymerization of cyclic monomers may become important in copolymerization systems. The four propagation reactions assumed to be irreversible in the derivation of the Mayo-Lewis equation must be modified to include reversible processes. Lowry[114, 115] first derived a copolymer composition equation for the case in which some of the propagation reactions are reversible and it was applied to ring-opening copolymerization systems[116, 117]. In the case of equilibrium copolymerization with complete reversibility, the following reactions must be considered.

$$M_1^* + M_1 \rightleftharpoons M_1 M_1^*$$
$$M_1^* + M_2 \rightleftharpoons M_1 M_2^*$$
$$M_2^* + M_1 \rightleftharpoons M_2 M_1^*$$
$$M_2^* + M_2 \rightleftharpoons M_2 M_2^*$$

A kinetic solution was obtained by Durgaryan[118], Ivin[119], Wittmer[120] and O'Driscoll[121]. A thermodynamic solution was obtained by Theil[122], Sawada[123], O'Driscoll[124, 125] and Leonard[126]. However, their application to a ring-opening system has not yet been published.

Another feature of the equilibrium copolymerization, shown in the next equations, is specific for ring-opening systems;

$$M_1 M_1^* + M_2 \rightleftharpoons M_1 M_2^* + M_1$$
$$M_2 M_2^* + M_1 \rightleftharpoons M_2 M_1^* + M_2$$

Such an exchange reaction occurs when the monomers attack the penultimate unit of the chain end, which is possible only for the case of cyclic propagating species. In the anionic copolymerization of lactams, this exchange reaction is faster than the propagation reaction and the copolymer composition is determined by this reaction and not by the propagation reaction[127]. A general solution of the copolymerization problem considering this equilibrium has not as yet been obtained.

From the above considerations, it is apparent that the copolymer composition is determined by the four propagation reactions only in limited cases. Thus, the apparent monomer reactivity ratios obtained by the application of the Mayo-Lewis equation to ring-opening polymerization systems may have no chemical meaning without confirming the validity of the elementary reactions. Furthermore, the rate constants may change with monomer feed composition due to a solvent effect and the existence of various propagating species must also be taken into consideration[30]. In the case of random copolymerization, the relative reactivity of monomers represents the ease with which monomer attack on the propagating species takes place. Thus, basicity or nucleophilicity of the monomer is a dominant factor in cationic copolymerization and electrophilicity of the monomer is the most important factor in anionic copolymerization[128].

b) Random Copolymers by Cationic Copolymerization

Cationic copolymerization of cyclic ethers, formals, esters and anhydrides has been thoroughly studied in recent years and sufficient information about it is now available. The propagating species involved in the cationic copolymerization of these oxacyclic monomers are believed to be the oxonium ions in most cases, but their detailed nature is dependent on monomer structure. From their copolymerization behavior, these monomers can be arranged in the following order of increasing carbocationic character of the propagating species:

| THF | PL | BCMO | IBO | DOL | St |

Monomer pairs close to each other in this series can form random copolymers, but remote monomer pairs are apt to yield homopolymer mixtures because of the dissimilarity of the propagating species.

Cationic copolymerization of oxacyclic monomers with vinyl monomers will be considered at first[129]. Styrene does not copolymerize with cyclic ethers or cyclic esters which propagate through the oxonium ions[130]. A typical example was the polymerization of α-methylstyrene with PL in which the formation of a homopolymer mixture occurred[98]. From studies on solvent effects in a similar copolymerization system, styrene and PL, the following scheme was suggested: the initiating species formed by styrene and the Lewis acid are effective only for the polymerization of styrene and those formed by the Lewis acid complex of PL are effective only for the polymerization of PL[131].

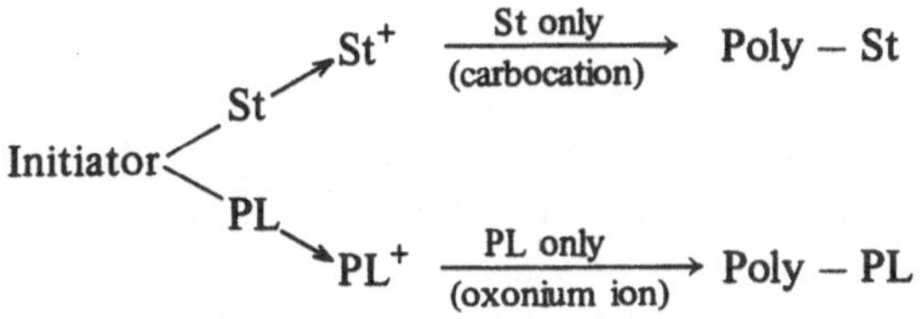

$$
\text{Initiator}
\begin{cases}
\text{St} \longrightarrow \text{St}^+ \xrightarrow[\text{(carbocation)}]{\text{St only}} \text{Poly} - \text{St} \\
\text{PL} \longrightarrow \text{PL}^+ \xrightarrow[\text{(oxonium ion)}]{\text{PL only}} \text{Poly} - \text{PL}
\end{cases}
$$

The propagating species involved in the polymerization of cyclic formal seem to resemble carbocations, and random copolymers are formed in the copolymerization of cyclic formals with styrene. For the copolymerization of DOL with styrene, the DOL-St cross-sequence was estimated, by NMR or by chemical methods, from the decrease of the formal unit in the copolymer and the formation of nearly random copolymer was confirmed[132].

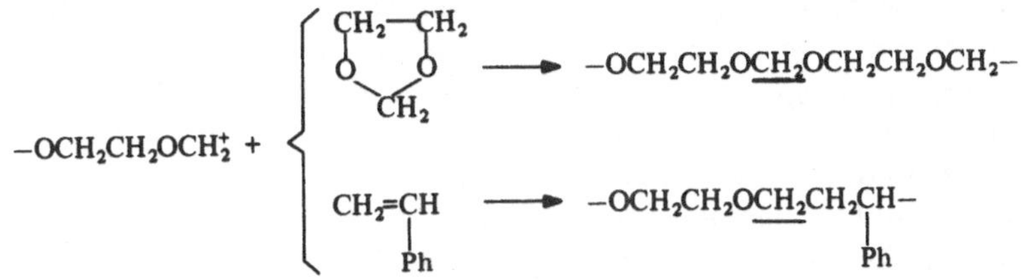

$$
-OCH_2CH_2OCH_2^+ +
\begin{cases}
\begin{array}{c} CH_2{-}CH_2 \\ O \quad\; O \\ CH_2 \end{array} \longrightarrow -OCH_2CH_2\underline{OCH_2}OCH_2CH_2OCH_2- \\[2em]
\begin{array}{c} CH_2{=}CH \\ | \\ Ph \end{array} \longrightarrow -OCH_2CH_2OCH_2\underline{CH_2CH}- \\ | \\ Ph
\end{cases}
$$

Several mechanisms were proposed for the propagation reaction in the cationic polymerization of cyclic formals[133]. The ring-expansion mechanism involving the propagation through a secondary oxonium ion was proposed by Plesch and supported by the cyclic structure of oligomers and polymers obtained from 1,3-dioxacycloalkanes[134] and by the ethoxy end-capping method[135]. However, cyclic oligomers are usually formed together with linear polymers[76, 136] and the propagation through nucleophilic attack on the living cationic end, proposed by Jaacks[137], should also be considered. Although both the $S_N 2$ mechanism through a tertiary oxonium ion[137] and the $S_N 1$ mechanism through a carbocation[138] can be considered, recent studies on the cationic polymerization of bicyclic formals suggested predominant occurrence of stereospecific $S_N 2$ attack at lower temperature and of non-stereospecific $S_N 1$ attack at higher temperature or in a polar solvent. This leads to the loss of stereospecificity and to random copolymerization with styrene[139].

In the cationic copolymerization of DOL with styrene considerable cleavage of polymer chains occurs if the styrene content is high but a molecular weight as high as 500,000 can be obtained if the styrene content is low[140]. In the case of trioxane and styrene, although cross-sequences were observed, the copolymerization system became heterogeneous and chain cleavage made the system very complex[141, 142]. The reactivity of isoprene in cationic polymerizations is similar to that of styrene, and copolymerization of isoprene with DOL yields random copolymers containing *trans*-1.4 polyisoprene units. The molecular weights of these polymers exceed 100,000 in some cases[143]. Vinyl ethers do not yield random copolymers even with DOL[144].

In contrast to cyclic formals, cyclic ethers do not yield random copolymers with styrene. Copolymerizations of tetrahydrofuran (THF) and 3,3-bischloromethyloxetane (BCMO) with styrene by Lewis acids gave only homopolymer mixtures[145]. This was also true for the reaction of N-vinylcarbazole with oxetane[97]. Studies on the copolymerization behavior of epoxides and styrene have clarified that the initial formation of polyether was accompanied by the polymer transfer with the polystyryl cation and produced apparent copolymers[112]. Ketene behaves similarly to styrene and has been shown to give random copolymers with DOL, epichlorohydrin and benzaldehyde, but not with cyclic ethers having more than four ring-atoms, thus distinguishing $S_N 1$-like monomers from $S_N 2$ monomers[148].

Cationic copolymerization of oxacyclic monomers has been thoroughly studied to clarify structure-reactivity relationships[130]. With these monomers random copolymers are formed rather easily. The randomness of copolymers derived from cyclic monomers has sometimes been demonstrated by comparing dyad concentrations determined by NMR with the calculated values based on apparent monomer reactivity ratios. In general, the agreement was fair. Figure 2 shows the results obtained in studies of isobutylene oxide-DOL copolymers[147]. It is clear that a random copolymer which obeys the Mayo-Lewis equation was formed in this case. Similar agreement was observed in the case of copolymerization of β,β-dimethyl-β-propiolactone with DOL or BCMO[148]. However, block character was observed in some cases. The formation of homopolymer mixtures in the copolymerization of THF and DOL[99] is a typical example and special care must be taken when we discuss copolymerizations involving different classes of monomers. Two factors are responsible for the block character. The first factor originates from the different nature of the propagating species and cyclic formals of enhanced carbocationic character are sometimes difficult to copolymerize with other oxacyclic monomers, which propagate through the oxonium ion. The second factor comes from the presence of different sites in a

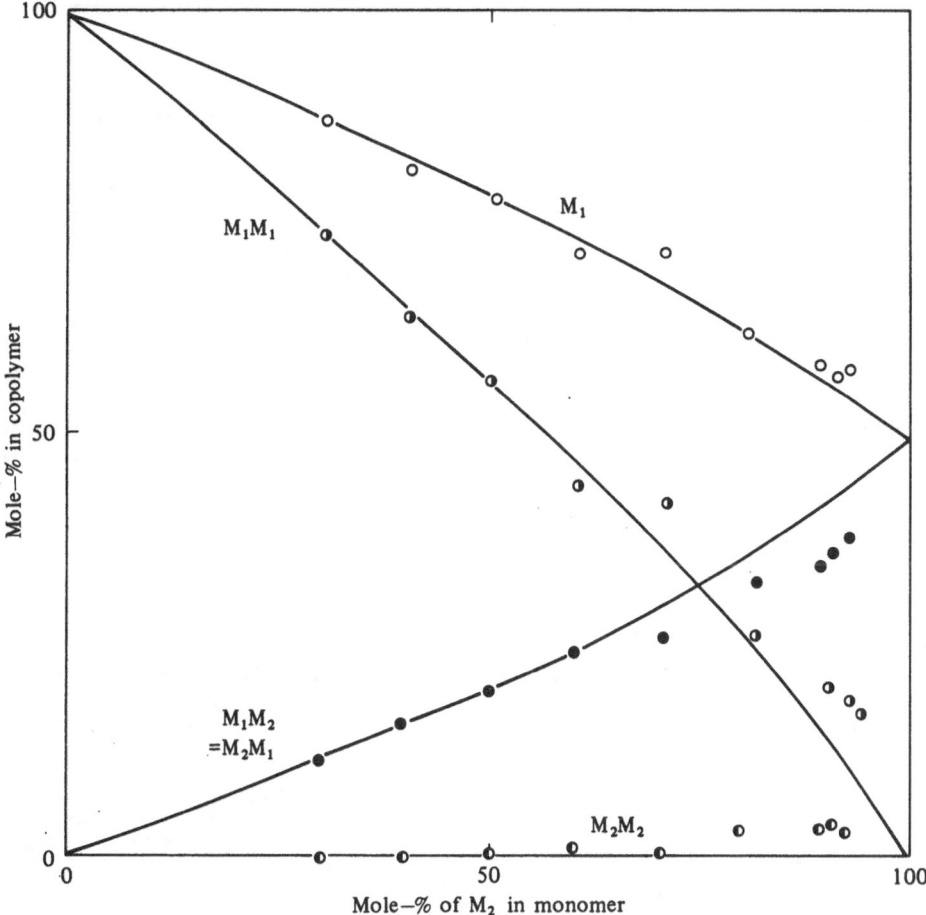

Fig. 2. Copolymer composition and diad concentrations for the IBO (M_1) and DOL (M_2) system. ○: Mole-% of M_1, ◐: mole-% of M_1M_1 diad, ●: mole-% of M_1M_2 or M_2M_1 diad, ◑: mole-% of M_2M_2 diad. Solid lines are theoretical

catalyst system. Epoxides and lactones polymerize easily by a coordinated type catalyst and the apparent copolymer might be a mixture of several different polymers formed from different sites.

The relative reactivities of cyclic ethers, formals and esters in the cationic copolymerization have been discussed in terms of the basicity of the monomers[130]. In the case of cyclic ethers, the reactivity obeys the following order of basicity: oxetane > THF > BCMO. Though less basic than other cyclic ethers, epoxides are highly reactive due to a ring strain. It was suggested that quantitative correlations of the copolymerization parameters with the basicity and the ring strain of the monomers exist among restricted monomer pairs[149, 150]. Basicity as a measure of nucleophilicity is usually expressed in terms of the ease of hydrogen bond formation[128, 130]. Recent studies on the kinetics of the cationic polymerization of cyclic ethers have demonstrated that the rate of homopolymerization was mainly

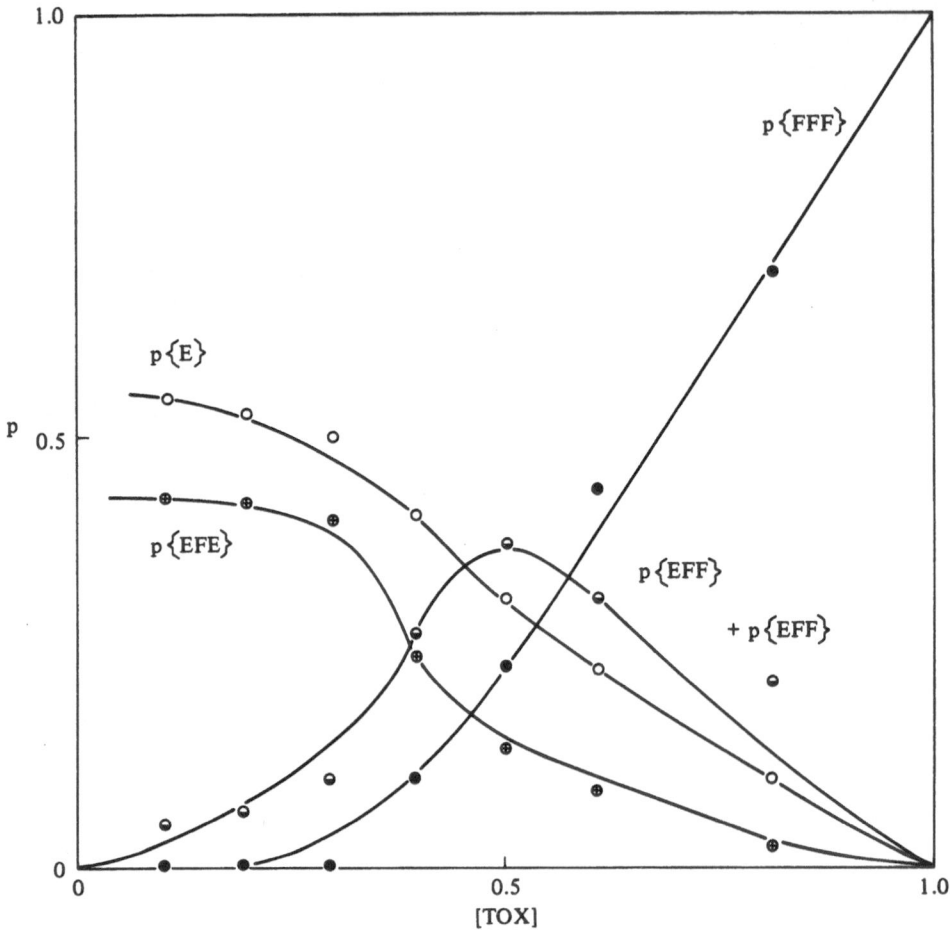

Fig. 3. Microstructure of copolymers of TOX and DOL as a function of the feed ratio

determined by the ease of ring-opening of the propagating oxonium ion and by the entropy decrease of the transition state[151]. Relative reactivity in homopolymerization is quite different from that in copolymerization.

Cationic copolymerization of THF is of interest in relation to the production of polyethers in the urethane industry. Although THF yields living polymers in cationic systems, copolymerization of THF and BCMO[241] or ε-caprolactone[52] proceeds with some termination or chain transfer reactions. Copolymerization of propylene oxide and tetrahydrofuran is usually carried out with a boron fluoride catalyst in the presence of glycol. Polyether glycols are produced together with cyclic tetramers[152, 153]. It is meaningless to discuss the relative reactivity without considering these oligomers[113]. Cationic copolymerization of trioxane (TOX) has been extensively studied to improve the thermal stability of polyacetal resin. Copolymerization of TOX with ethylene oxide was studied[154, 155, 156, 157], and NMR analysis of the polymerization system[158] and the pyrolysis gas chromatogram of the copolymer[159] clarified the mechanism. Copolymerization of TOX with DOL was also studied[159–164] and the microstructure of the copolymer indicated the complex nature of the polymerization system[111]. The copolymer composition versus triad fraction diagram shown in Fig. 3 indicates that TOX units incorporated in the copolymer chain appear as separate formal units $-OCH_2-$ rather than as three successive formal units $-(OCH_2)_3-$. The more reactive DOL becomes less reactive with increasing dilution of the solution because of its higher equilibrium concentration. Further characteristics of the polyacetal are the rearrangement of copolymer sequences by transacetalization[165]. Cyclic oligomers soluble in methanol can often be detected. It is necessary to characterize all of the products so as to understand the mechanism of the reaction[166].

Cationic copolymerization of the monomers, which do not polymerize by themselves, has been studied in recent years. Monomers in this category are five-and six-membered oxacyclic monomers such as tetrahydropyran[166, 167], 1,4-dioxane[166, 167], γ-butyrolactone[168] and cyclic anhydrides[169]. Effective comonomers for the copolymerization are epoxides and oxetanes. The enhanced cationic reactivity of epoxides and oxetanes is caused by ring strain. Frequently, a tendency to produce alternating copolymers is observed. This will be discussed separately. Cationic copolymerization of THF with cyclic anhydrides[170, 171] was also studied kinetically[172].

The effect of propagation-depropagation equilibrium on the copolymer composition is important in some cases. In extreme cases, depolymerization and equilibration of the heterochain copolymers become so important that the copolymer composition is no longer determined by the propagation reactions. Transacetalization, for example, cannot be neglected in the later stages of trioxane and DOL copolymerization[111, 173]. This reaction is used in the commercial production of polyacetal in which redistribution of acetal sequences increases the thermal stability of the copolymers.

In the copolymerization of five- and six-membered oxacyclic monomers, the effective monomer concentration in the propagation reaction decreases because only the monomer in excess of equilibrium is available for copolymerization. However, it is not easy to determine the equilibrium monomer concentration in a copolymerization system. The following equilibrium is expected to exist in the copolymerization of THF.

The first equilibrium was indicated by a decrease of THF segment lengths in the copolymer at higher temperature and at higher dilution[116, 117]. In Fig. 4, the effect of temperature on the cationic copolymerization of THF and BCMO is shown[117]. This is explained by the easy depropagation of the oxonium ion of THF, which has a penultimate THF unit. The second equilibrium is not important. A recent kinetic study on the polymerization of THF has shown that the endo-attack at the electrodeficient carbon atom in the ring occurred thirty times faster than the exo-attack at the penultimate carbon atom in the chain[174].

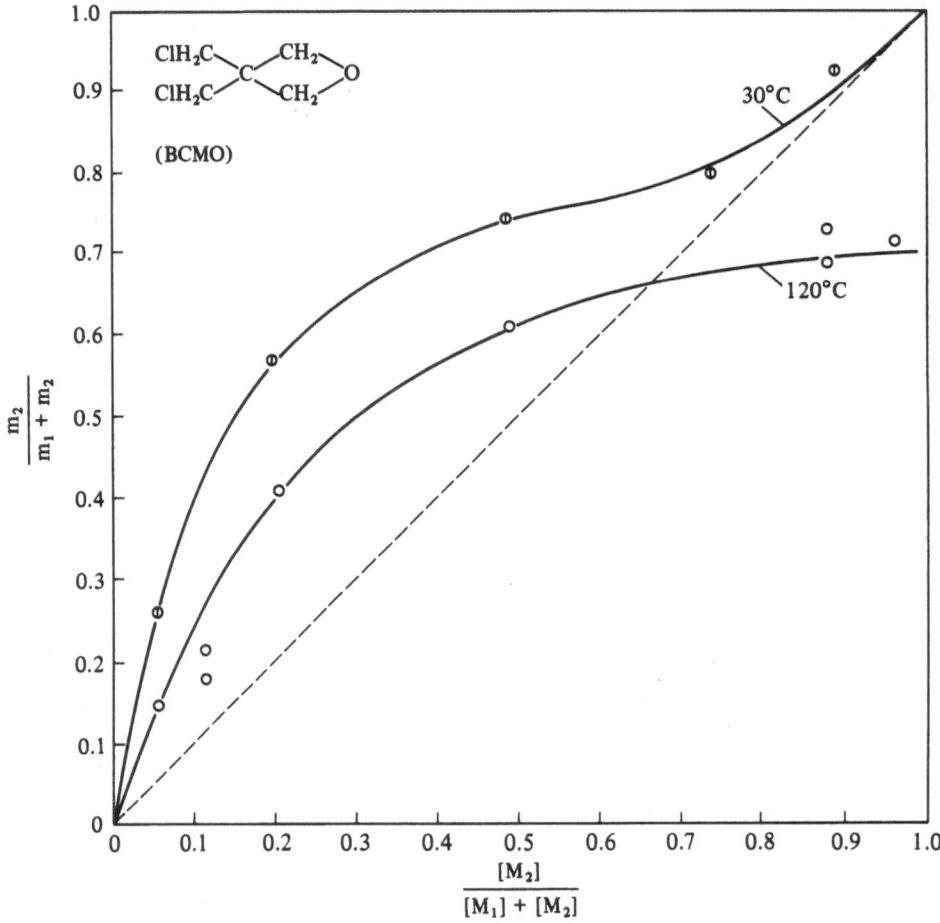

Fig. 4. Dependence of copolymer composition on monomer feed in copolymerization of BCMO with THF at 30 and 120 °C

The effect of a catalyst is important in cationic copolymerizations. Epoxides and β-lactones form random copolymers only with trialkyl aluminum catalysts. Unusual sequence distributions were observed in the cationic copolymerization of epoxides or lactones using Lewis acids[175−177] and have been attributed to the diversity of catalyst species. The formation of ferric alkoxide by the reaction of ferric chloride and propylene oxide was well established[178] and triethyloxonium tetrafluoroborate was synthesized by the reaction of boron fluoride etherate and epichlorohydrin[179]. In cationic polymerizations employing Lewis acid catalysts, the termination is caused by the reaction of the propagating cation with its gegenanion, while the metal halide gradually changes to metal alkoxide[47]. Schaeffer et al. concluded that dual catalyst sites are present during the cationic copolymerization of epoxides with sulfur dioxide or maleic anhydride[180, 181]. Tada et al. suggested that denaturation of a boron fluoride catalyst occurs during the copolymerization of PL and BCMO[182]. Thus, it is reasonable to assume that the Lewis acid is modified by epoxides and lactones to form coordinate anionic or coordinate cationic catalysts which favor the polymerization of these monomers[183]. The existence of long sequences in the PL-BCMO copolymer became apparent from NMR[183] and crystallographic[184] studies. The polymerization of lactones is more favored by coordinate cationic sites than by free cationic sites. The reactivity of γ-butyrolactone in the copolymerization with BCMO increases in the order $HClO_4 < BF_3 \cdot Et_2O <$ $< SnCl_4$[185]. Catalysts based on boron fluoride or the triethylaluminum-water system had quite a different effect on the composition of copolymers obtained from the copolymerization of cyclic ethers and lactones[186–188]. The presence of coordination sites in triethylaluminum-water catalysts was demonstrated by the formation of crystalline polyepichlorohydrin[189]. Recent crystallographic studies on aluminum catalysts favor the coordinate cationic mechanism[190]. The active species of the triethylaluminum water catalyst were shown to be Et_2 AlOAlEt$_2$ and $(EtAlO)$[191], and stereospecific polymerization of propylene oxide with a crystalline binuclear organometallic catalyst was found to proceed through coordination of the monomer to the catalyst[192, 193]. However, the measured rate constant for THF polymerization was essentially the same, being independent of whether triethylaluminum catalysts were used[194] but the coordinate cationic mechanism remains to be clarified.

Cationic copolymerization of other cyclic monomers has been studied less extensively. Such copolymerization between substituted 2-oxazolines has been reported[195].

The relative rate of cationic homopolymerization is governed by three factors, *i.e.* the concentration of the propagating species, the ring-opening reactivity of the growing species and the nucleophilic reactivity of the monomer. From kinetic studies[196, 197] of the polymerization of oxazolines and oxazines it was found that the second factor was the most important. On the other hand, the relative reactivity in the cationic copolymerization is mainly determined by the nucleophilicity of the monomer and for 2-substituted 2-oxazolines this is in the order of benzyl > methyl > > isopropyl > H > phenyl[195].

c) Random Copolymers by Anionic and Coordinate Copolymerization

Anionic copolymerization of cyclic monomers occurs only between similar monomer pairs. Random copolymers are not formed between vinyl monomers and epoxides or lactones[198] because the propagating species are very different. Thus, the success of the copolymerization of cyclic disulfide and nitropropylene was an exceptional case[199].

Anionic copolymerization involving similar monomer pairs which propagate through a similar chain end occurs rather easily. Examples are the copolymerization involving aldehydes, isocyanate and ketenes, although these are not always random copolymerizations[200–205]. The copolymerization of isothiocyanate and episulfides has been reported[206]. Anionic copolymerization of ethylene oxide and propylene oxide easily takes place and electrodeficient ethylene oxide is the more reactive comonomer. NMR analysis of the microstructure of this copolymer has been reported[207]. Copolymerization between episulfides also occurs easily and dyad or triad sequences have been determined by NMR[208, 209].

Anionic copolymerization of lactams presents an interesting example of copolymerization. Studies of the copolymerization of α-pyrrolidone and ε-caprolactam showed that α-pyrrolidone was several times more reactive than ε-caprolactam at 70 °C, but became less reactive at higher temperatures due to depropagation[210, 211]. By analyzing the elementary reactions Vofsi *et al.*[127] concluded that transacylation at the chain end occurred faster than propagation and that the copolymer composition was essentially determined by the transacylation equilibrium and the acid-base equilibrium of the monomer anion together with the usual four elementary reactions of the copolymerization.

$$AcPy + Cl^- \rightleftharpoons AcCl + Py^-$$
$$Py^- + CLH \rightleftharpoons PyH + CL^-$$
$$AcPy + Py^- \longrightarrow AcPyPy^-$$
$$AcPy + CL^- \longrightarrow AcPyCL^-$$
$$AcCL + Py^- \longrightarrow AcCLPy^-$$
$$AcCL + CL^- \longrightarrow AcCLCL^-$$

Anionic copolymerization of ε-caprolactam and ω-caprylolactam was also reported[212, 213]. Organosiloxane copolymers can be prepared from two different cyclics by using acid or base catalysts[214].

Extensive studies of stereoselective polymerization of epoxides were carried out by Tsuruta *et al.*[215]. Copolymerization of a racemic mixture of propylene oxide with a diethylzinc-methanol catalyst yielded a crystalline polymer, which was resolved into optically active polymers[216, 217]. Asymmetric selective polymerization of d-propylene oxide from a racemic mixture occurs with asymmetric catalysts such as diethyzinc- (+) borneol[218]. This reaction is explained by the asymmetric adsorption of monomers onto the enantiomorphic catalyst site[219]. Furukawa[220] compared the selectivities of asymmetric catalysts composed of diethylzinc amino acid combinations and attributed the selectivity to the bulkiness of the substituents in the amino acid. With propylene sulfide, excellent asymmetric selective polymerization was observed with a catalyst consisting of diethylzinc and a tertiary-butyl substituted α-glycol[221, 222].

Copolymerizations of α-amino acid NCA's have been studied but the results are less conclusive because of solvent effects. Essentially random copolymers were obtained in poor solvents like acetonitrile and block type copolymers were obtained in moderately good solvents like DMF or nitrobenzene[224]. Copolymerization of mixtures of L and D α-amino acid NCA's revealed that the enantiomer being in excess of the starting monomer was preferentially incorporated into the copolymer[225–227]. The cause of the asymmetric selection was ascribed to the chirality of the growing polymer chains[215]. Copolymerization of racemic mixtures of α-amino acid NCA's under specified conditions gives stereoregular polymers. The preferred initiator systems are primary amines[228, 229], quaternary ammonium compounds[230], alkali metal compounds[231], trialkyl aluminum[232, 233] and transition metal compounds[234, 237]. Asymmetric selective polymerization was also studied by using chiral initiators[232, 234, 235, 236].

d) Alternating Copolymers

Alternating copolymers may be considered new homopolymers having specific structure units. Synthetic chemists have been interested in controlling the copolymer sequences and preparing designed copolymers at their will. Although few alternating copolymers of sufficiently high molecular weight have been prepared by ring-opening polymerization, such polymers are interesting from the standpoint of the mechanism of polymerization. Two mechanisms are conceivable for the formation of alternating copolymers. Monomers which have potential reactivity toward electrophilic or nucleophilic attack but cannot polymerize by themselves are apt to form alternating copolymers. Such monomers are carbon monoxide, carbon dioxide, sulfur dioxide, cyclic acid anhydride and carbonyl compounds etc. An alternative mechanism is the polymerization through a complex formed between comonomers. Charge transfer complexes are formed between a donor monomer and an acceptor monomer and they appear to be the active species in radical copolymerization. In ring-opening polymerization systems, spontaneous polymerization of a betaine type complex was found quite recently[265].

Copolymers of carbon monoxide, carbon dioxide, sulfur dioxide or carbon disulfide are frequently formed in combination with oxiranes, thiiranes or aziridines. Copolymerization of carbon monoxide and ethylenimine was carried out under radiation and the formation of 3-nylon was observed[238].

$$CO\ +\ \overset{CH_2-CH_2}{\underset{NH}{\diagdown\diagup}}\ \longrightarrow\ -CONHCH_2CH_2-$$

In the ethylene atmosphere, carbon monoxide and ethylenimine copolymerized with a radical initiator to form a terpolymer[239]. The following radical mechanism may be proposed:

$$R + CH_2=CH_2 \longrightarrow R-CH_2CH_2\bullet \overset{CO}{\longrightarrow} R-CH_2CH_2CO\bullet \overset{\overset{CH_2-CH_2}{\underset{NH}{\diagdown\diagup}}}{\longrightarrow}$$

$$RCH_2CH_2CONHCH_2CH_2\bullet \longrightarrow -(CH_2CH_2CONH)_{\overline{n}}CH_2CH_2-$$

The formation of polyesters from carbon monoxide and propylene oxide using a cobalt catalyst may involve an alternate coordination on the metal and an insertion of monomers into the carbon-transition metal bond[240].

$$CO + CH_3CH-CH_2 \xrightarrow{\ Co(acac)_3-AlEt_3\ } \overset{CH_3}{-OCHCH_2CO-}$$

Incorporation of carbon dioxide as a reactive comonomer has been studied by several groups. Inoue *et al.* were the first to succeed in preparing high molecular weight polycarbonates by the copolymerization of carbon dioxide and propylene oxide[241, 242];

$$CO_2 + \underset{O}{CH_2-CH-CH_3} \xrightarrow{ZnEt_2-H_2O} -CH_2CH(CH_3)OCOO^-$$

The mechanism of this alternating copolymerization was explained by the following equation:

$$-OCOOZn + \underset{O}{CH_2-CH-CH_3} \longrightarrow -OCOOCH_2CH(CH_3)OZn$$

$$-CH_2CH(CH_3)OZn + CO_2 \longrightarrow -CH_2CH(CH_3)OCOOZn$$

Zinc carbonate reacts with epoxide to form zinc alkoxide, which in turn reacts with carbon dioxide to regenerate zinc carbonate. The most effective catalyst systems were the reaction products between diethylzinc and polyhydroxy compounds such as water or polyhydric phenols[243, 244]. This copolymer is interesting as a biodegradable elastomer[245].

Aziridines can be used to obtain partial polyurethane units from carbon dioxide. Various catalysts including cationic systems[246-249] were proposed.

$$CO_2 + \underset{\underset{R'}{N}}{CH_2-CH-R} \longrightarrow -OCONR'CHRCH_2-$$

Saegusa *et al.* succeeded in preparing alternating oligomers from ethylene phenylphosphonite, acrylic monomer and carbon dioxide through zwitter ions[250].

X = CN, COOCH$_3$

Yamazaki *et al.* prepared polyureas and polythioureas from carbon dioxide and disulfide with diamines *via* the N-phosphonium salts[251].

Carbon disulfide yielded oligomeric polythiocarbonates by copolymerization with episulfides[252].

Cationic copolymerization of sulfur dioxide and propylene oxide was studied and the product was identified as polysulfite ethers[180, 253];

$$\text{SO}_2 + \underset{\underset{O}{\diagdown\diagup}}{\overset{\text{CH}_2-\text{CH}-\text{CH}_3}{}} \longrightarrow +\text{CH(CH}_3)\text{CH}_2\text{O}]_n\text{SOO}-$$

Cationic copolymerization of other monomers which do not polymerize by themselves often yields alternating copolymers. Some examples are[254, 255]:

$$\text{PhCHO} + \text{CH}_2\text{=CHPh} \xrightarrow{\text{BF}_3\text{·Et}_2\text{O}} \underset{\underset{\text{Ph}}{|}}{-\text{O}-\text{CH}}-\text{CH}_2-\underset{\underset{\text{Ph}}{|}}{\text{CH}}-$$

and

$$\text{PhCHO} + \text{CH}_2\text{=C=O} \xrightarrow{\text{BF}_3\text{·Et}_2\text{O}} \underset{\underset{\text{Ph}}{|}}{-\text{O}-\text{CH}}-\text{CH}_2-\underset{\underset{\text{O}}{\|}}{\text{C}}-$$

Apparently alternating copolymers were obtained from the copolymerization of 2-methyl tetrahydrofuran-BCMO[186], tetrahydropyran-BCMO[186], γ-butyrolactone-BCMO[168], BCMO-THF[187, 188], 3,3-dimethyl and 3,3-bisfluoromethyl oxetane-THF[117] and succinic anhydride-THF[256], but it is necessary to determine the co-monomer distributions in these copolymers to discuss the alternating tendency. In some cases, significant influence of the catalyst system on copolymer composition was reported[168, 186–188]. Recently Hsieh[257] reported the formation of alternating terpolymers consisting of epoxide, acid anhydride and THF, but the exact mechanism remains unsolved.

Anionic copolymerization of monomers which do not polymerize by themselves sometimes yields alternating copolymers.

$$\text{PhCHO} + \text{CH}_2\text{=C=O} \xrightarrow{\text{BuLi}} \underset{\underset{\text{Ph}}{|}}{-\text{CH}}-\text{O}-\underset{\underset{\text{O}}{\|}}{\text{C}}-\text{CH}_2-$$

$$\text{PhCHO} + (\text{CH}_3)_2\text{C=C=O} \xrightarrow{\text{BuLi–ZnEt}_2} \underset{\underset{\text{Ph}}{|}}{-\text{CH}}-\text{O}-\underset{\underset{\text{O}}{\|}}{\text{C}}-\underset{\underset{\text{CH}_3}{\overset{\text{CH}_3}{|}}}{\text{C}}-$$

Benzaldehyde and ketene form alternating copolymers with both cationic and anionic catalysts[255]. In the case of dimethylketene, crystalline alternating copolymers of benzaldehyde or acetone were obtained with a stereospecific catalyst[258]. The pioneering work of Natta can be interpreted in terms of an alternate coordination of monomers on the catalyst. Even acetophenone and methyl formate gave crystalline copolymers of dimethylketene with an organometallic catalyst[259, 260].

Cyclic acid anhydrides are highly reactive not only to electrophilic attack but also to nucleophilic attack. Alternating copolyesters were obtained from cyclic acid anhydrides and epoxides at high temperatures[261−263].

$$
\underset{\text{CO}}{\overset{\text{CO}}{\diagdown}}\text{O} + \text{CH}_2\text{–CH–CH}_2\text{Cl} \xrightarrow{\text{LiCl}} \cdots -\text{CO} \quad \text{COOCHCH}_2\text{O–} \atop \text{CH}_2\text{Cl}
$$

The preferred catalysts are salts of inorganic and organic acids as well as tertiary amines. Phthalic anhydride, succinic anhydride and maleic anhydride are typical acid anhydrides, while ethylene oxide, propylene oxide, epichlorohydrin and phenyl glycidyl ether are typical epoxides. The synthesis of a ladder polymer was carried out by using bisanhydrides[264].

$$
\text{CH}_2\text{–CHCH}_2\text{OPh} + \text{(bisanhydride)} \xrightarrow{R_3N} \text{(ladder polymer)}
$$

With cationic catalysts, epoxides become so reactive that polyesterethers are formed instead of polyesters[256].

Spontaneous polymerization through intermediate zwitter ions was extensively studies by Saegusa[265]. A monomer with nucleophilic reactivity (M_N) is mixed with a second monomer (M_E) with electrophilic reactivity. The zwitter ion is formed spontaneously at room temperature in dipolar aprotic solvents. Reaction between the two zwitter ions gives an oligomeric zwitter ion which acts as an initiating species. Propagation consists of successive addition of the zwitter ion to the macrozwitter ion and also the intermolecular reaction between two oligomeric or polymeric zwitter ions. This idea was illustrated by alternating copolymerization of 2-oxazoline with β-propiolactone[266, 267].

$$
\text{2Z} \longrightarrow \cdots \longrightarrow +\text{CH}_2\text{CH}_2\text{N–CH}_2\text{CH}_2\text{CO}+_n \atop \quad\quad\quad\quad \text{HCO} \quad\quad \text{O}
$$

Several other types of monomers have been successfully used for the spontaneous copolymerization[266]. For M_N, 2-oxazine, N-benzyliminotetrahydrofuran, N-benzylidene aniline, dioxaphospholanes and 1,3,3-trimethylazetidine were used and as M_E, succinic anhydride, propane sultone, acrylic acid and acrylamide were used[268–273].

Ethylenimine is known to spontaneously form copolymers with β-propiolactone[274] or cyclic imides[275]. Crystalline copolyamides of alternating structure were obtained with the latter monomer.

Propanesultone also forms random and alternating copolymers with N-substituted ethylenimines[276] and similar mechanisms involving zwitter ions were suggested. The polymerization through zwitter ions was performed with cyclic phosphonite as the M_N monomer and pyruvic acid as the M_E monomer[250].

Polymerization of a stable intermediate prepared from two monomers can produce an alternating copolymer. The amino acid azide hydrobromide method and the amino acid succinimidyl ester hydromide method were applied to prepare ordered copolyamides[277] and sequential copolypeptides[278–280].

$$HBr \cdot NH_2CH_2CH_2CONH(CH_2)_5CON_3 \xrightarrow[DMF]{Et_3N} \{NH(CH_2)_2CONH(CH_2)_5CO\}_n$$

4. Block and Graft Copolymers

Significant advances have been made in the synthesis, characterization and utilization of block and graft copolymers over the last ten years[281, 282]. A major contribution to these developments has been the living polymer technique pioneered by
Szwarc[283]. The preparation of polymers with well defined structure by this technique made it possible to clarify fundamental characteristics of block and graft copolymers such as domain formation and thermoplastic elasticity. The living tendency
frequently observed in ring-opening polymerization provides novel methods for the
preparation of tailor-made polymers with specific properties, such as surface activity.

a) Syntheses of Prepolymers with Functional Groups

Living polymers of styrene and α-methylstyrene possess very reactive carbanions
at one or both of their ends and their molecular weight distributions are narrow.
Vacuum line techniques are necessary to prepare them. Cumyl potassium, butyl
lithium and the disodium salts of α-methylstyrene tetramer in tetrahydrofuran are
used as initiators to prepare mono- and bifunctional carbanions. The conversion of
living anions to other functional groups has been studied and quantitative conversion
to carboxyl groups and hydroxyl groups was achieved by reactions with carbon
dioxide and ethylene oxide[284, 285]. Living polymerization techniques were also
applied to butadiene and isoprene, while the carbanion at the polymer ends, was
converted to carboxyl or hydroxyl groups[286, 287].

Special care must be taken for the preparation of living polymers from methyl
methacrylate and vinylpyridine because of the inevitably occurring side reactions
at room temperature. Ring-opening polymerization has also been used for the synthesis of functional prepolymers. Hydroxyl-terminated polyethers and polyesters
of low molecular weight are employed in the polyurethane industry. Base-catalyzed
ring-opening polymerizations of ethylene oxide and propylene oxide are of primary
importance and successive addition of ε-caprolactone to a hydroxylic initiator is also
used. Cationic polymerization of tetrahydrofuran is used for the preparation of
poly(tetramethylene glycol). The hydroxyl groups can be chemically modified to
carboxyl groups, chloroformate groups or isocyanate groups by reaction with cyclic
acid anhydrides, phosgene or diisocyanates.

Of recent interest is the living cation of tetrahydrofuran. The oxonium ion with
suitable counter ions is stable and exists in equilibrium with the monomer[47]. Bifunctional initiators were found to produce the polytetrahydrofuran dication.
Octamethylene bis-1,3-dioxolenium perchlorate

$$
\begin{array}{c}
CH_2-O \\
\quad\quad\quad > C^+-(CH_2)_8-C^+ < \quad\quad\quad \\
CH_2-O \quad\quad\quad\quad\quad O-CH_2 \\
\quad\quad\quad\quad\quad\quad\quad O-CH_2
\end{array}
\quad \bullet \ 2ClO_4^- \quad [288]
$$

diacyl cation $R(CO^+)_2 \cdot 2SbF_6^-$ [289] and pyrosulphuryl fluoride $(FSO_2)_2O$ [290] were
bifunctional initiators. The dication of polytetrahydrofuran reacts with ammonia
or potassium cyanate to give di-primary amine or diisocyanate[290].

Linear polysiloxanes containing terminal functional groups such as alkoxy groups, chlorine atoms are technically prepared by equilibration of cyclic polysiloxanes with functional silicone compounds[291].

Prepolymers of vinyl monomers with functional end groups are also prepared by free radical polymerization. Azo initiators with carboxyl or hydroxyl groups are also prepared by free radical polymerization. Azo initiators with carboxyl or hydroxyl groups are successfully used for this purpose. Disulfides are effective chain transfer agents for preparing prepolymers with functional end groups. Telechelic polymers of dienes have been extensively studied for use as liquid rubbers[292, 293]. Various types of polycondensation and polyaddition reactions also can be used to obtain prepolymers with terminal functional groups.

Graft copolymers are usually prepared from copolymers whose backbone attaches functional groups which can be converted into grafting sites. A variety of techniques for synthesizing copolymers with backbone grafts have been investigated[294].

b) Block and Graft Copolymers by Successive Addition

For the successive addition of cyclic monomers to a macromolecular initiator the following factors must be considered: purity of the macromolecular initiator with suitable functional groups, the rate of the initiation reaction and the initiator efficiency; and the living tendency of the polymerization. If the terminal functional groups in the macromolecular initiator are less pure or the living tendency of the system is low due to accompanying chain transfer reactions, the copolymer obtained will be a mixture of a block copolymer and a homopolymer. When initiation by the macromolecular initiator is not satisfactory because of the slow initiation relative to the propagation reaction or because of the non-bonding initiation with transfer, the initiator efficiency will be lowered and the copolymer obtained will have a broad distribution of compositions.

Anionic polymerization of ethylene oxide by living carbanions of polystyrene was first carried out by Szwarc[295]. A limited number of methods have been reported in the preparation of A—B and A—B—A copolymers in which B was polystyrene and A was poly(oxyethylene)[296—298]. The actual procedure was to allow ethylene oxide to polymerize in a vacuum system at 70 °C with the polystyrene anion initiated with cumyl potassium in THF[299]. The yields of pure block copolymers are usually limited to about 80% because homopolymers are formed[300].

$$-CH_2-CH^- \quad + \quad CH_2-CH_2 \quad \longrightarrow \quad -CH_2-CH-CH_2-CH_2-O^-$$
$$\underset{Ph}{|} \qquad\qquad \underset{O}{\diagdown\diagup} \qquad\qquad\qquad\qquad \underset{Ph}{|}$$

In a similar manner polyisoprene-polyethylene oxide block copolymers can be prepared[301]. It is surprising that the poly(methyl methacrylate) anion can be successfully used for the polymerization of ethylene oxide without chain transfer[302]. Graft copolymers are also prepared by successive addition of ethylene oxide to the polystyrene backbone metalized with potassium at the diphenylmethane units[303]. Block and graft copolymers of poly(oxyethylene) are soluble in both water and benzene.

Another method consits of the anionic polymerization of vinyl monomers by alkali metal salts of poly(oxyethylene) glycol. Block copolymers of methyl methacrylate or acrylonitrile and ethylene oxide are obtained with an insufficient yield, because the initiator efficiency is low even in polar solvents[304, 305]. Addition of crown ethers greatly improved on the situation[306]. Free radical polymerization of vinyl monomers can be used to prepare block copolymers from poly(oxyethylene) prepolymers. Mechanical stirring of methyl methacrylate solution of poly(ethylene oxide) generates free radicals which initiate the polymerization of MMA, but the conversion is not high[307]. Macromolecular initiators derived from poly(oxyethylene) glycol have been used for the polymerization of vinyl monomers. Tobolsky developed a unique method which involved the introduction of peroxy end groups into prepolymers by two step reactions with diisocyanate and hydroperoxide[308]. An alternative procedure consisted of trichloroacetylation of the end groups and redox initiation with sulfur dioxide and pyridine[309]. In these two methods the block efficiency was not high. An improved method was proposed by Furukawa in which a macromolecular initiator is prepared from polyetherglycol and azobiscyanovaleroyl dichloride[310, 311].

Graft copolymers of ethylene oxide were produced by successive addition to cellulose, starch and nylon in the presence of alkali metal catalysts[312−314]. Partial titration of hydroxylic polymers in dimethylsulfoxide with sodium-naphthalene in tetrahydrofuran increases the grafting efficiency by promoting the metalation[313]. However, this reaction yielded only graft copolymers containing short ethylene oxide segments.

Cationic polymerization of ethylene oxide is accompained by depolymerization and oligomerization. It has been reported that ethylene oxide polymerized cationically with the living dication of tetrahydrofuran and a surface active material was obtained[290].

Block copolymers of propylene oxide and ethylene oxide were the earliest examples of the commercial production of sequential anionic block copolymers[316]. The hydrophobic poly(propylene oxide) segment connected with the hydrophilic polyethylene oxide segment exhibits a remarkable surface activity. Block copolymers having a propylene oxide sequence are useful as segmented elastomers[281]. Segmented polyurethanes are prepared from polyacetal glycols. Radical polymerization of vinyl monomers is a useful procedure to obtain block copolymers from prepolymers with peroxy end groups or polyazo groups[308−311].

The cationic addition of THF appears to be a new technique for the preparation of block and graft copolymers. Saegusa et al.[317] obtained a block copolymer of THF and 3,3-bischloromethyloxetane (BCMO) by the polymerization of BCMO with the living polymer of THF. By a similar procedure an A−B−A′ type block copolymer consisting of (poly-THF)-(THF/BCMO random copolymer)-(poly-BCMO) was prepared and found to exhibit thermoplastic elasticity[317]. The procedure employed is shown schematically in Fig. 5. Macromolecular initiators containing dioxolenium groups at the chain ends of poly-THF and polystyrene have been used to initiate the polymerization of cyclic ethers and formals[318]. Although THF formed

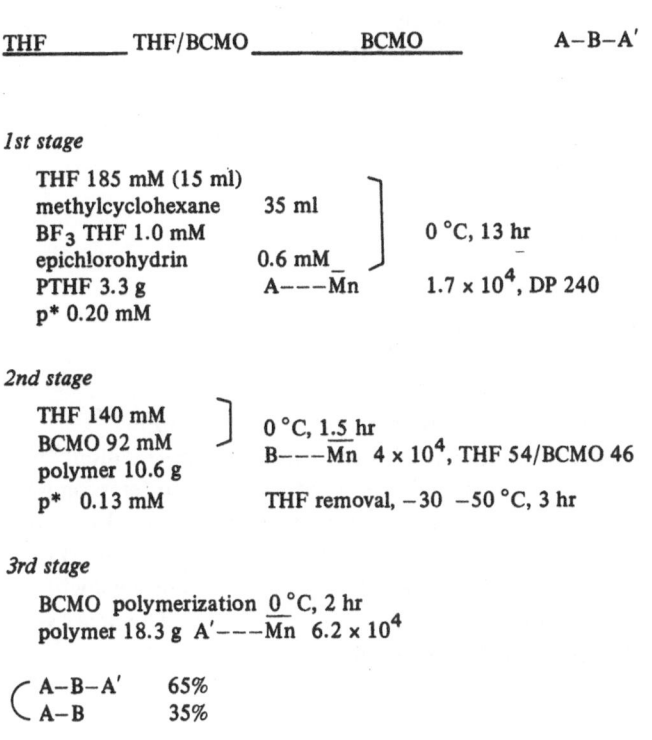

THF _____ THF/BCMO _____ BCMO _____ A–B–A′

1st stage

THF 185 mM (15 ml)
methylcyclohexane   35 ml
BF$_3$ THF 1.0 mM
epichlorohydrin   0.6 mM   0 °C, 13 hr
PTHF 3.3 g   A———$\overline{M}$n   1.7 × 10^4, DP 240
p* 0.20 mM

2nd stage

THF 140 mM   0 °C, 1.5 hr
BCMO 92 mM   B———$\overline{M}$n 4 × 10^4, THF 54/BCMO 46
polymer 10.6 g
p* 0.13 mM   THF removal, −30 −50 °C, 3 hr

3rd stage

BCMO polymerization 0 °C, 2 hr
polymer 18.3 g A′———$\overline{M}$n 6.2 × 10^4

A–B–A′   65%
A–B   35%

Fig. 5. Scheme of the three-stage block copolymerization of THF

a living chain from the initiator, BCMO yielded a block copolymer and termination reactions were encountered. 7-Oxa-bicyclo(2.2.1)-heptane yielded a mixture of homopolymer and block copolymer. This fact indicates the occurrence of a chain transfer reaction. Polymerization of dioxolane and tetraoxane by a polymeric initiator yielded no block copolymer, demonstrating that, in the case of cyclic formals, a different mechanism is involved in the initiation reaction. This similar behavior was encountered in the polymerization of BCMO by the living polymer of 1,3-dioxolane, where only a mixture of homopolymer and block copolymer was obtained[319].

Block or graft copolymers can be obtained by cationic polymerization of THF with macromolecular initiators. The recommended groups for the initiation are the dioxolenium cation, the acyl cation and the super acid ester, each of which can be introduced into the backbone polymer by reaction with silver salts of strong acids. Introduction of the dioxolenium group into polystyrene was carried out by the following route[320]:

$-CH_2-CH-$ $\quad$ $\begin{array}{c}\text{1. } CH_3COCl \\ \hline \text{2. } NaOBr\end{array}$ $\quad$ $-CH_2-CH-$ $\quad$ $\begin{array}{c}\text{3. } SOCl_2 \\ \hline \text{4. } HOCH_2CH_2Br\end{array}$

(with COOH substituent)

$-CH_2-CH-$ $\quad$ $\begin{array}{c}AgClO_4 \\ \hline C_3H_7NO_2\end{array}$ $\quad$ $-CH_2-CH-$

(COOCH$_2$CH$_2$Br substituent) → (cyclic $O-C-O^+ \cdot ClO_4^-$)

At first an acyl cation was used together with an unfavorable gegenanion[319, 320]. Although the acetyl cation gave an inferior result even with SbF_6^- gegenanion[101], recent investigations showed that a higher acyl cation with SbF_6^- counteranion gives more satisfactory results[323, 324]. Reactive halogen groups in polyvinyl chloride or chlorinated SBR are allowed to react with silver triflate in the THF solution to form graft copolymers[325, 326]. Unexpectedly, the reaction mixture became a gel when the graft copolymerization was continued for several hours[320, 324]. This was explained by the formation of a polymeric oxonium ion.

$-CH_2-CH-CH_2-\overset{CH_3}{\underset{COCl}{C}}-$ $\xrightarrow[THF]{AgSbF_6}$ $-CH_2-CH-CH_2-\overset{CH_3}{\underset{CO^+ \cdot SbF_6^-}{C}}-$ $\xrightarrow{THF}$ $-CH_2-CH-CH_2-\overset{CH_3}{\underset{COO(CH_2)_4-O^+}{C}}-$ $\cdot SbF_6^-$

$-CH_2-CH-CH_2-\overset{R}{\underset{Cl}{\underset{\,}{C}}}-$ (with Cl) $\xrightarrow[THF]{AgOSO_2CF_3}$ $-CH_2-CH-CH_2-\overset{R}{\underset{OSO_2CF_3}{C}}-$ (with Cl) $\xrightarrow{THF}$

$\longrightarrow$ $-CH_2-CH-CH_2-\overset{R}{\underset{O(CH_2)_4-O^+}{C}}-$ (with Cl) $\cdot OSO_2CF_3^-$

Although a considerable number of publications describe the addition of aldehydes to living anions and claim the formation of block copolymers, the evidence for successful block polymerization is not convincing[327-329]. The polymerization must be performed in vacuum to minimize contamination by impurities and the end capping of the polymers at low temperatures is necessary for successful work. Block efficiency is high in the case of chloral polymerization where a living polymer is formed[330], but not sufficiently high in the case of β-cyanopropinaldehyde[331]. Isocyanates can also be incorporated into block copolymers by anionic processes. Polymer anions from styrene, isoprene and methyl methacrylate were used as initiators for the polymerization of butyl isocyanate[332]. 2,4-Tolylene diisocyanate polymerized with these initiators and gave soluble block copolymers for which the following structure was suggested:

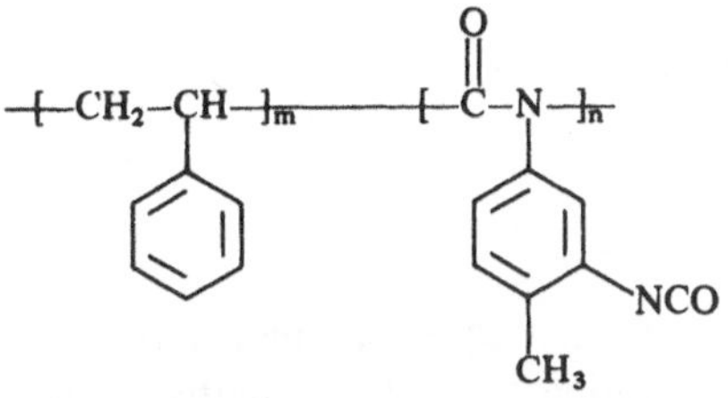

Anionic grafting of β-propiolactone onto acrylate copolymers was extensively studied[333−335].

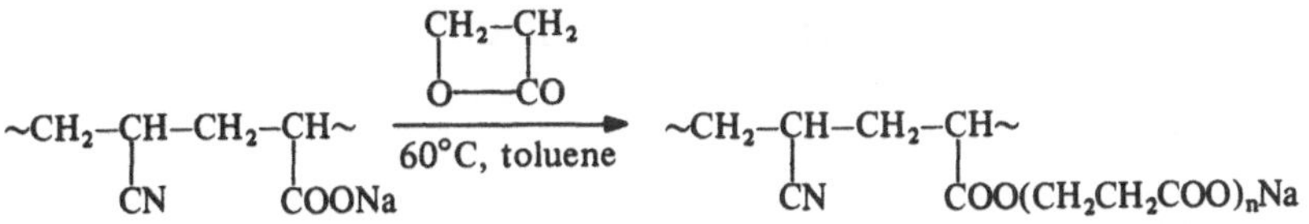

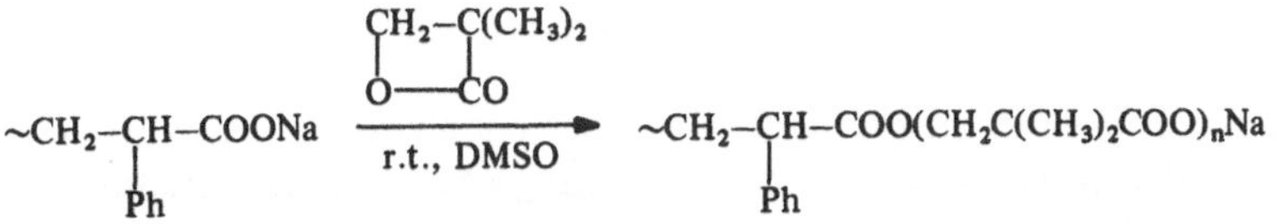

Anionic polymerization of pivalolactone with the polystyrene anion produced only homopolymer mixtures, but the polystyrene carboxylate anion was able to give a block copolymer[336]. The block efficiency depends on catalyst ratio and conversion because the initiation step is slow compared with propagation[337]. Tough and elastic films were obtained by graft copolymerization or block copolymerization of pivalolactone onto elastomers containing tetrabutylammonium carboxylate groups[338, 339].

Addition of δ- and ε-lactones to the hydroxyl group of a polymer end occurs at high temperature[340]. With the copolymer of styrene and 2-hydroxyethyl methacrylate used as a backbone chain, grafting of ε-caprolactone at 200 °C with stannous octoate produced polymeric plasticizers[341]. Anionic addition of ε-caprolactone to a living polymer anion at lower temperature is accompanied by cyclic oligomer formation if the polymerization is continued for long. Block copolymers were prepared by successive addition of ε-caprolactone to the living anions of polystyrene, polyisoprene, poly(methyl methacrylate) and poly(oxy ethylene)[342−344]. Anionic block copolymers of styrene and butadiene prepared by sec-butyl lithium were used as initiators for the polymerization of ε-caprolactone and yielded elastomers compatible with various polymers[345].

Anionic polymerization of alkylene sulfides was extensively studied by Sigwalt and collaborators. In contrast to propylene oxide, anionic polymerization of pro-

$$-CH_2-CH-CH_2-\overset{\overset{\displaystyle CH_3}{|}}{\underset{\underset{\displaystyle COOCH_2CH_2O[CO(CH_2)_5O]_nH}{|}}{C}}-$$

$$C_4H_9-(CH_2-CH-)_{\overline{m}}-(CH_2-CH=CH-CH_2-)_{\overline{n}}-[CO(CH_2)_5O-]_{\overline{p}}-H$$

pylene sulfide yielded living polymers. Living polymers of styrene, α-methylstyrene[346-348], butadiene, isoprene[349, 350], vinylpyridine[351, 352] and methyl methacrylate[346] were used as initiators. The initiation mechanism follows the equations[353]:

$$RNa + \underset{S}{\overset{\triangle}{CH_2-CH-CH_3}} \longrightarrow RSNa + CH_2=CH-CH_3$$

$$RSNa + \underset{S}{\overset{\triangle}{CH_2-CH-CH_3}} \longrightarrow RSCH_2\overset{\overset{\displaystyle CH_3}{|}}{CH}SNa$$

This procedure was applied to the synthesis of triblock copolymers ABA of thiiranes by coupling the AB block copolymers or by using a bifunctional initiator[348]. Thietanes consisting of four membered cyclic sulfides differ from thiiranes in the nature of the propagating species. The attack of carbanion on thietane was found to occur at the sulfur atoms and the new carbanion formed by ring opening was the propagating species[353].

Thermoplastic elastomers of the ABA type with poly(alkylene sulfide) as the hard segment A were prepared by stepwise addition of ethylene sulfide, isobutylene sulfide, or thietane to the poly(propylene sulfide) dianion[346, 352, 354, 355] or the polydiene dianion[356-358]. ABA block copolymers, prepared by the sequential polymerization of styrene, isoprene and ethylene sulfide, were strong resilient elastomers[357]. Successive addition of ethylene sulfide and isobutylene sulfide to the polyoxyethylene anion was attempted with the hope of obtaining block copolymers but chain transfer occurred in the reaction of alkoxide end groups of living polyoxyethylene with propylene sulfide, thus block copolymers were not obtained[359]. When the polytetrahydrofuran cation was added to propylene sulfide and thietane formation of some block copolymers was observed[360].

Anionic polymerization of lactams was shown to proceed according to what is called the activated monomer mechanism. With bischloroformates of hydroxy-terminated poly(tetramethyleneglycol) and poly(styrene glycol) as precursors for a polymeric initiator containing N-acyl lactam ends, block copolymers with α-pyrrolidone and ϵ-caprolactam were obtained by bulk polymerizations in vacuum at 30 and 80 °C, respectively[361].

$$RCON \left< \begin{matrix} (CH_2)_n \\ CO \end{matrix} \right) + Na^+ \; ^-N \left< \begin{matrix} (CH_2)_n \\ CO \end{matrix} \right) \rightarrow RCON(CH_2)_nCON \left< \begin{matrix} (CH_2)_n \\ CO \end{matrix} \right)$$
$$\underset{Na}{|}$$

$$HN \left< \begin{matrix} (CH_2)_n \\ CO \end{matrix} \right)$$
$$\xrightarrow{} RCONH(CH_2)_nCON \left< \begin{matrix} (CH_2)_n \\ CO \end{matrix} \right) + NaN \left< \begin{matrix} (CH_2)_n \\ CO \end{matrix} \right)$$

A polymeric initiator for the anionic polymerization of α-pyrrolidone can be prepared by radical polymerization of styrene with 4,4'-azobis(pyrrolidone 4-cyanovalerate)[362]. It became evident that degradation of poly(ε-caprolactam) was important at the later stage of the block copolymerization, especially at high temperatures[361]. In fact, fractionation of nylon-6 block copolymers, obtained from tolylene diisocyanate terminated polybutadiene or polystyrene, has shown that contamination with ε-capprolactam homopolymer was unavoidable in the polymerization at 160 °C[363]. Esters are weak cocatalysts (initiators) for the anionic polymerization of lactams and acrylate copolymers can be used with ε-caprolactam to obtain graft copolymers[364]. Polyaddition of lactams to poly(butylenediamine) with acid catalysts at high temperature gave flexible polyamide-polyether block copolymers[365]. The formation of graft copolymers improved on the compatibility of polypropylene and nylon-6[366]. The reaction takes place at the melt temperature between polypropylene, treated with maleic anhydride, and the amine ends of nylon-6.

Oxazolines are among the few monomers which yield cationic living polymers. An interesting block copolymer consisting of poly(N-lauroyl ethyleneimine) and poly(N-acetyl trimethyleneimine) was prepared[367] and found to be of use as a surfactant.

Polybutadiene with the tosylate end groups was used as an initiator for the polymerization of 2-oxazoline[368], but block efficiency was not high because of slow initiation[369]. The resulting copolymer was hydrolyzed and butadiene-ethylenimine copolymers, soluble in chloroform[368], were obtained.

$$HO-(CH_2-CH-)_m-OH \xrightarrow{TsCl} TsO-(CH_2-CH-)_m-OTs \longrightarrow$$
$$\quad\quad\quad |CH{=}CH_2 \quad\quad\quad\quad\quad\quad\quad |CH{=}CH_2$$

$$\longrightarrow HO-(CH_2CH_2N-)_l-(CH_2-CH-)_m-(NCH_2CH_2-)_n-OH$$
$$\quad\quad\quad\quad\quad\quad |HCO \quad\quad |CH{=}CH_2 \;\; |HCO$$

$$\xrightarrow{OH^-} HO-(CH_2CH_2NH-)_l-(CH_2CH-)_m-(NHCH_2CH_2-)_n-OH$$
$$\quad\quad\quad\quad\quad\quad\quad\quad |CH{=}CH_2$$

Grafting of 2-methyloxazoline onto chloromethylated polystyrene beads in benzonitrile at 110 °C gave graft copolymers, which were hydrolyzed to poly(styrene-g-ethylenimine) and useful as a chelating resin[370].

Sequential polypeptides have been of interest to biochemists and polymer chemists. The formation of living polymers from N-carboxy-α-amino acid anhydrides (NCA) was confirmed by using ^{14}C labeled primary amine initiators[371]. However, it was suggested that the chain transfer reaction or the termination reaction may occur depending upon the nature of the initiators and solvents used for the polymerization[372]. When the polymerization of NCA was initiated with n-butylamine in acetonitrile, the amino end group in the precipitated polymer retained an activity as a polymerization initiator[373]. Block copolypeptides of various chain lengths consisting of different α-amino acids were synthesized by polymerization of the corresponding NCA in acetonitrile with the precipitated polypeptide, which has been formed by polymerizing an NCA with n-butyl amine in acetonitrile[374]. Block copolymers of styrene and γ-methyl-D-glutamate were obtained by polymerization of the NCA with a primary amine terminated polystyrene. Higher molecular weights were obtained in ethyl acetate than in propionitrile or dimethyl formamide[375].

Recently, Gallot et al. succeeded in preparing well characterized block copolymers by polymerization of γ-benzyl-L-glutamate NCA and ε-carbobenzoxyl-L-lysine NCA with a primary amine ended polystyrene or polybutadiene in dioxane or ben-

$$CH_2{=}CH + NH_2-\!\!\bigcirc\!\!-S{-}S{-}\!\!\bigcirc\!\!-NH_2 \xrightarrow{AIBN} NH_2-\!\!\bigcirc\!\!-S{-}(PSt){-}S{-}\!\!\bigcirc\!\!-NH_2$$
$$\quad |PH$$

$$NH_2-\!\!\bigcirc\!\!-S{-}(PSt){-}S{-}\!\!\bigcirc\!\!-NH_2 \xrightarrow[NH_2NH_2]{} NH_2CH_2CONH-\!\!\bigcirc\!\!-S{-}(PSt){-}S{-}\!\!\bigcirc\!\!-NHCOCH_2NH_2$$

$$NH_2-(PSt)-NH_2 + \quad \underset{NH-CO}{\overset{CH_2CH_2COOCH_3}{\overset{|}{CH-CO}}}\!\!\!\!\!\bigg\rangle O \quad \longrightarrow \quad NH_2[(CHCONH)-]_n-PSt-[(NHCOCH)]_nNH_2$$
$$\quad\quad\quad\quad\quad\quad\quad\quad\quad\quad\quad\quad\quad CH_2CH_2COOCH_3 \quad CH_2CH_2COOCH_3$$

zene[376, 377]. End capping of the carboxyl group with hexamethylene diamine was employed for the synthesis of the prepolymers.

The mechanism of the polymerization of NCA with tertiary amine is still controversial. Mori and Iwatsuki claim that the true initiator is the primary amino group formed by hydrolysis of the NCA with contaminated water and that tertiary amine forms a complex with the NCA and accelerates the addition reaction[378]. Harwood *et al.* confirmed the propagating carbamate by NMR in polymerization initiated with a strong base[379]. The successive addition of NCA to the polymer end catalyzed with a strong base affords an alternative procedure for the synthesis of block copolypeptides. Block copolypeptides of poly(oxyethylene) were prepared by triethyl amine catalyzed polymerization of NCA in the presence of poly(oxyethylene) bischloroformate[380].

The synthesis of block copolymers containing polysiloxane segments has been studied by sequential polymerization and by coupling techniques. The resulting block copolymers showed low temperature flexibility, excellent electrical properties, durability toward weathering and high degrees of gas permeability. Living polystyrene potassium in tetrahydrofuran was used to initiate the polymerization of octamethylcyclotetrasiloxane at room temperature[381]. Living polymers of isoprene or methyl methacrylate have also been used[382]. The conversion of cyclosiloxane to block copolymer was low because the initially formed silanolate led to an equilibration of polysiloxane[383]. Recently, rapid polymerization of cyclotrisiloxanes and slow equilibration of lithium silanolate were discovered and the block efficiency for this sequential polymerization has increased substantially.

$$C_4H_9(CH_2CH)_nLi \xrightarrow{m \; [-Si(CH_3)_2O-]_3} C_4H_9(CH_2CH)_n(Si-O)_{3m}Li$$

Lithium ended polystyrene was used for the polymerization of hexamethylcyclotrisiloxane in tetrahydrofuran or diglyme as a promoting solvent[384-388]. Foam fractionation revealed some distribution of polydimethylsiloxane segments in the block copolymer[389]. Polystyryl potassium was changed to polystyryl lithium by adding lithium tetraphenylborate in THF for use in the successive addition of cyclotrisiloxane[390]. Sequential anionic polymerization of different cyclotrisiloxanes using lithium compounds as initiators to produce different siloxane block copolymers. The following example shows preparation of a block copolymer containing soft dimethylsiloxane and hard diphenyl siloxane segments[391]:

$$LiO\underset{Ph}{\overset{Ph}{Si}}OLi \xrightarrow[THF]{[-Si(CH_3)_2O-]_3} LiO\underset{CH_3}{\overset{CH_3}{Si}}O \sim O\underset{CH_3}{\overset{CH_3}{Si}}OLi$$

$$\xrightarrow{[-Si(Ph)_2O-]_3} LiO\left(\underset{Ph}{\overset{Ph}{Si}}O\right)_m\left(\underset{CH_3}{\overset{CH_3}{Si}}O\right)_n\left(\underset{Ph}{\overset{Ph}{Si}}O\right)_m Li$$

c) Block and Graft Copolymers by Reaction Between Polymers

It is possible to prepare designed block copolymers by coupling two different pre-polymers. The reactivity of functional groups at the polymer end is assumed to be equivalent to the functional groups of low molecular weight compounds. When an attempt is made at preparing a block copolymer by coupling prepolymers with functional groups at their ends, the following factors must be considered: purity of the functional end groups, molar ratio of the reagents, rate of the coupling reaction, absence of side reactions and the selection of appropriate solvents. Purity of the functional end groups is sufficiently high when the living polymer technique is employed, while it is low in the free radical methods, where 80 ~ 90% purity is usual and there is a decrease in block efficiency. The molar ratio of the reagents is important in the coupling reaction, and stoichiometry is required to obtain high yields of high molecular weight block copolymers[392]. Fast coupling reactions are very important to obtain well-defined block copolymers because the concentration of the functional groups is very low. Although many attempts at preparing block copolymers have been made, only moderate success was achieved by the reaction between NH_2 or the OH group and COCl or NCO group[393]. It is worth mentioning that care must be taken to avoid side reactions such as those caused by contaminating water. The reaction between the functional groups at the ends of a polymer chain may be retarded by slow diffusion, but this has not as yet been confirmed. Segregation or phase separation in the concentrated polymer solution may also decrease the rate of coupling between the functional groups at the ends of different polymers even in an apparently homogeneous solution.

Coupling reaction of living polymers has been exhaustively investigated with respect to the synthesis of ABA block elastomers from AB block copolymers[281]. Carbanions at the end of the polystyrene or polyisoprene chain may be linked together with a variety of coupling agents such as dibromoalkane, bischloromethyl ether, diethyl sebacate and carbon dioxide, but phosgene or dimethyl dichlorosilane appears to be more effective. Quantitative coupling occurs more easily with the alkoxide or thiolate anion. Phosgene was used to prepare ABA block copolymers from the α-methylstyrene-propylene sulfide block anion[348].

$$C_2H_5(CH_2-\underset{\underset{C_6H_5}{|}}{\overset{\overset{CH_3}{|}}{C}}\longrightarrow)_m(CH_2-\underset{\underset{}{}}{\overset{\overset{CH_3}{|}}{CH}}-S)_n-Li \xrightarrow{COCl_2} \alpha MS-PS-\alpha MS$$

The gel permeation chromatogram shown in Fig. 6 illustrates the purity of a block copolymer obtained by ion coupling. It is seen that about 5% of uncoupled block copolymer contaminates a triblock copolymer of narrow molecular weight distribution. The synthesis of star block polymers owes its recent development to the use of new coupling agents[412].

Ion coupling of anionic and cationic living polymers is an interesting procedure for the synthesis of a well-defined block copolymer. Attempted coupling of the polystyrene anion with the poly-THF cation initiated by triethyloxonium tetrafluoroborate yielded a block copolymer mixed with homopolymers[394]. The block ef-

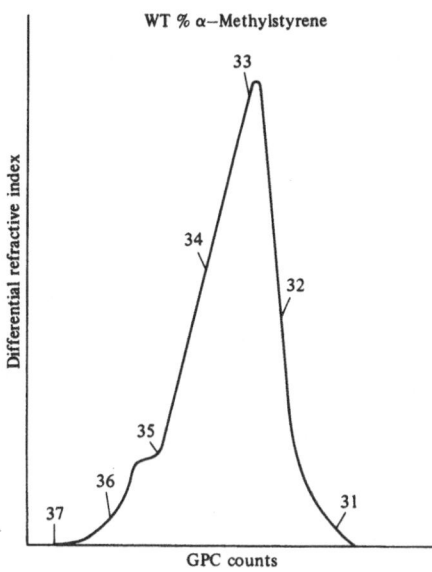

Fig. 6. Gel permeation chromatogram of the MS-PS-MS block copolymers obtained by coupling

ficiency was much improved by changing the initiator for poly-THF to super acid derivatives[290].

$$\sim OC_4H_8 \quad \text{(cyclic)} \quad \bullet \ FSO_3^- \ + \ \sim CH_2\text{--}\overset{-}{C}H\ Na^+ \longrightarrow \sim OC_4H_8OC_4H_8\overset{|}{C}HCH_2\sim$$
$$\underset{Ph}{} \qquad\qquad\qquad \underset{Ph}{}$$

Further improvements were made by carboxylating the polystyrene anion, leading to quantitative yields[395, 396]. Multiblock copolymers of molecular weight 500,000 consisting of sixty segments of poly-THF and polystyrene were prepared[397]. The effect of the molar feed ratio of the two components is shown in Fig. 7.

$$\sim OC_4H_8 \quad \text{(cyclic)} \quad \bullet \ ClO_4^- \ + \ \sim CH_2\overset{|}{C}HCOO^- Na^+ \longrightarrow \sim OC_4H_8OC_4H_8OCO\overset{|}{C}HCH_2\sim$$
$$\underset{Ph}{} \qquad\qquad\qquad\qquad \underset{Ph}{}$$

Block copolymers containing polysiloxane segments are of great interest as polymeric surfactants and elastomers. Polycondensation and polyaddition reactions of functionally ended prepolymers are usually employed to prepare well-defined block copolymers. The living polystyrene anion reacts with α,ω-dichloropoly(dimethylsiloxane) to form multiblock copolymers[398].

$$\text{NaStNa} \ + \ Cl(\overset{\overset{\displaystyle CH_3}{|}}{\underset{\underset{\displaystyle CH_3}{|}}{Si}}O)_n\overset{\overset{\displaystyle CH_3}{|}}{\underset{\underset{\displaystyle CH_3}{|}}{Si}}\text{--}Cl \longrightarrow [St\text{--}Siloxane]_m$$

An alternative procedure involved the coupling of polysilanolate, obtained by sequential addition of cyclotrisiloxane to the polystyrene anion, with α,ω-diacetoxysiloxane[385, 386, 399].

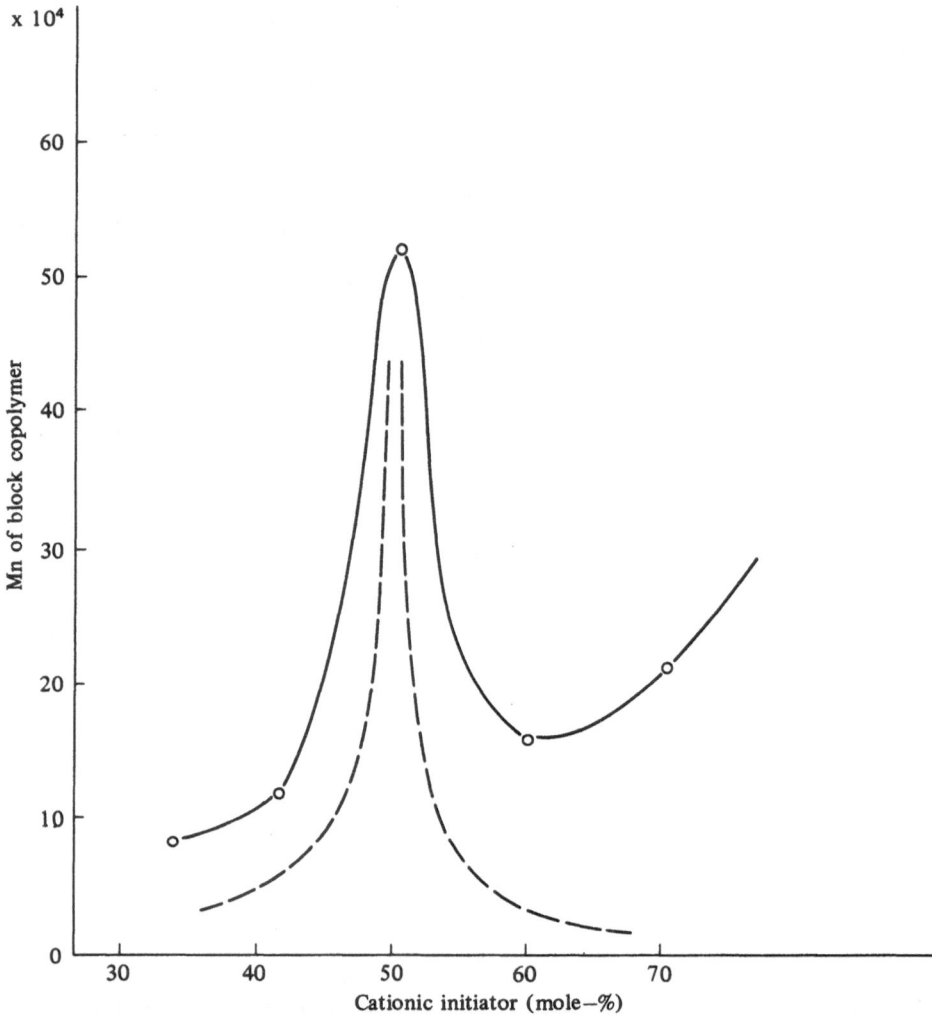

Fig. 7. Effect of the molar ratio of initiators on the $\overline{\text{Mn}}$ of the block copolymer (THF-MS$_4$)n
produced by ion coupling
————— observed   – – – – calculated

The reaction between a hydroxy ended prepolymer and α, ω-dichloropoly(dimethylsiloxane) has been studied[400]. It was found that coupling with N,N-dimethyl-
aminopolysiloxane prepared from chlorosiloxane was preferable[401–406].

$$\text{POH} + \text{Me}_2\text{N}\underset{\overset{\displaystyle |}{\underset{\displaystyle \text{CH}_3}{}}}{\overset{\overset{\displaystyle \text{CH}_3}{\displaystyle |}}{\text{Si}}}\text{-O-R} \longrightarrow \text{P-O-}\underset{\overset{\displaystyle |}{\underset{\displaystyle \text{CH}_3}{}}}{\overset{\overset{\displaystyle \text{CH}_3}{\displaystyle |}}{\text{Si}}}\text{-O-R} + \text{Me}_2\text{NH}$$

Hydrosilylation was successfully applied to block copolymer synthesis. Polystyrene with two H–Si end groups derived from the polystyrene dianion and di-

methylchlorosilane was added to diallylpolysiloxane in the presence of platinum catalysts[398, 400]. Various types of block and graft copolymers have become available using this technique[407].

$$
\begin{array}{ccc}
\text{CH}_3 & \text{CH}_3 \\
| & | \\
\text{HSi}-\text{St}-\text{SiH} + \text{CH}_2{=}\text{CHCH}_2\text{SiO}-\text{OSiCH}_2\text{CH}{=}\text{CH}_2 & \xrightarrow[140°C]{\text{H}_2\text{PtCl}_6} & -\text{St}-\text{Si}-(\text{CH}_2)_3-\text{Si}-\text{O}- \\
| & | \\
\text{CH}_3 & \text{CH}_3
\end{array}
$$

Because of the interest in the polyurethane industry with respect to grafting, extensive studies were made on the preparation of polydimethyl siloxane-polyoxy-alkylene block copolymers by coupling of polyetherglycol and polysiloxanes with functional terminal groups[408].

Coupling reactions were also used for synthesizing graft copolymers. Only a few examples are quoted here. Grafting of the potassium alkoxide derivative of poly-(ethylene oxide) onto poly(methyl methacrylate) in toluene occurred at 110 °C and soluble graft copolymers were obtained[409, 410]. Polypropylene, reacting with maleic anhydride, was coupled with the amino end group of nylon[411].

Interchange reactions between two different polymers were observed in the case of polyester or polyamide, but this subject is beyond the scope of the present review.

5. Scope of Future Research

The mechanism of ring-opening polymerization has received much attention in recent years. The studies of it has made control of the polymerization reactions possible resulting in desirable products. However, many problems still remain unsolved in this field. In fact, the situation is far less satisfactory than the fields of vinyl polymerization and polycondensation. Extensive studies of copolymerization should be useful for the establishment of the chemistry of ring-opening polymerization.

Because of the great differences in the properties between vinyl polymers and heterochain polymers, copolymerization of a vinyl monomer and a cyclic monomer seems very intersting. Yet, little success has been achieved in the formation of random copolymers because the reactivities are very different between vinyl monomers and cyclic monomers. However, recent progress in the field of organic chemistry has suggested many possibilities especially for the activation of monomers and for the modification of the reactivity of the propagating species. The probability of successful synthesis of random copolymers has thus greatly increased.

Another interesting field is the utilization of unused resources such as carbon monoxide, carbon dioxide, etc. Development of useful copolymers is expected to come by applying the modern techniques of polymer chemistry. Ring-opening copolymerization should be one of the most likely methods for this purpose.

Preparation of block and graft copolymers in various applications is interesting if we consider the limited availability of new polymers. For the synthesis of tailor-made polymers, useful for fundamental research work, ring-opening polymerization has versatile applications. In addition to the living polymer technique of anionic polymerization of vinyl monomers, various types of new block and graft copolymers could be synthesized by ring-opening polymerization. Especially useful are the specific properties of heterochain polymers, which are not found in vinyl polymers.

6. References

[1] Dainton, F. D., Ivin, K. J.: Quart. Rev., *12*, 61 (1958)
[2] Saegusa, T., Shiota, T., Matsumoto, S., Fujii, H.: Polymer J. *3*, 40 (1972)
[3] Ivin, K. J.: Angew. Chem. *85*, 533 (1973)
[4] Wichterle, O., Stehlicek, J., Kodaira, T., Sebenda, J.: J. Polymer Sci. B *5*, 931 (1967)
[5] Sawada, H.: J. Macromol. Sci. C *5*, 151 (1970)
[6] Bonetskaya, A. K., Sukuratov, S. M.: Vysokomol. Soedin. A *11*, 532 (1969)
[7] Brown, H. C., Shechter, H.: J. Amer. Chem. Soc. *76*, 467 (1954)
[8] Wittbecker, E. L., Hall, H. K. Jr., Campball, T. W.: J. Amer. Chem. Soc. *82*, 1218 (1960)
[9] Hall, H. K. Jr.: J. Org. Chem. *28*, 233 (1963)
[10] Sumitomo, H., Hashimoto, K., Ando, M.: J. Polymer Sci., Polymer Lett. Ed. *11*, 631 (1973)
[11] Hall, H. K. Jr.: J. Amer. Chem. Soc. *80*, 6412 (1958)
[12] Sakai, S., Fujinami, T., Sakurai, S.: J. Polymer Sci., Polymer Lett. Ed. *11*, 631 (1973)
[13] Szwartz, T. D., Hall, H. K. Jr.: J. Amer. Chem. Soc. *93*, 137 (1971)
[14] Levy, A., Litt, M.: J. Polymer Sci. B *5*, 881 (1967)
    Bassiri, T. G., Levy, A., Litt, M.: J. Polymer Sci. B *5*, 871 (1967)
[15] Mukaiyama, T., Fujisawa, T., Nohira, H., Hyugaji, T.: J. Org. Chem. *27*, 3337 (1962)
[16] Szwarc, M.: Adv. Polymer Sci. *4*, 1 (1965)
[17] Ballard, D. G. H., Tighe, B. J.: J. Chem. Soc. B 702, 976 (1967)
[18] Miki, T., Higashimura, T.: J. Polymer Sci. A-1, *5*, 2997 (1967)
[19] Dachs, K., Schwartz, E.: Angew. Chem. *74*, 540 (1962)
[20] Goodman, I., Nesbitt, B. F.: Polymer *1*, 384 (1960)
[21] Vogl, O., Knight, A. C.: Macromolecules *1*, 315 (1968)
[22] Dreyfuss, M. P., Dreyfuss, P.: J. Polymer Sci. A-1, *4*, 2179 (1966)
[23] Mita, I., Imai, I., Kambe, H.: Makromol. Chem. *137*, 169 (1970)
[24] De Sokgo, M., Pepper, D. C., Szwarc, M.: Chem. Comm. *1973*, 419
[25] Plesch, P. H.: Polymer. *8*, 137 (1971)
[26] Saegusa, T., Hashimoto, Y., Matsumoto, S.: Macromolecules *4*, 1 (1971)
[27] Saegusa, T., Matsumoto, S.: Macromolecules *1*, 442 (1968)
[28] Goethals, E. J.: Polymer Preprints *13*, (1), 51 (1972)
[29] Saegusa, T., Ikeda, H., Fujii, H.: Macromolecules *5*, 359 (1972)
[30] Szwarc, M.: Carbanions, living polymers and electron transfer process. New York: Interscience 1968
[31] Boileau, S., Sigwalt, P.: Europ. Polymer J. *4*, 3 (1968)
[32] Kazanski, K. S., Solovyanov, A. A., Enteils, S. G.: Europ. Polymer J. *7*, 1421 (1971)
[33] Hall, H. K. Jr.: Macromolecules *2*, 488 (1969)
[34] Lundberg, R. D., Doty, P.: J. Amer. Chem. Soc. *79*, 3961 (1957)
[35] Fenn, D. J., Thomas, M. D., Tighe, B. J.: J. Chem. Soc. (B) 1044 (1970)
[36] Schwan, T. C., Price, C. C.: J. Polymer Sci. *40*, 457 (1959)
[37] Boileau, S., Sigwalt, P.: Europ. Polymer J. *3*, 57 (1967)
[38] Goethals, E. J.: J. Macromol. Sci., Chem. A *7*, 1375 (1973)
[39] Deffieux, A., Boileau, S.: ACS Polymer Preprints *18*, 699 (1977)
[40] Fessler, W. A., Juliano, P. C.: Ind. Eng. Chem. Product Res. Dev. *11*, 407 (1972)
[41] Sangster, J. M., Worsfold, D. J.: Macromolecules *5*, 229 (1972)

42) Kagiya, T., Matsuda, T., Hirata, R.: J. Macromol. Sci., Chem. A 6, 451 (1972)
43) Hemery, P., Boileau, S., Sigwalt, P.: J. Polymer Sci. C 52, 189 (1975)
44) Kobayashi, S., Danda, H., Saegusa, T.: Bull. Chem. Soc. Japan 47, 2699 (1974)
45) Matyjaszewski, K., Kubisa, P., Penczek, S.: J. Polymer Sci., Polymer Chem. Ed. 12, 1333 (1974)
46) Saegusa, T.: Makromol. Chem. 175, 1199 (1974)
47) Dreyfuss, P., Dreyfuss, M. P.: Adv. Polymer Sci. 4, 528 (1967)
48) Saegusa, T.: J. Macromol. Sci., Chem. A 6, 997 (1972)
49) Vofsi, D., Tobolsky, A. V.: J. Polymer Sci. A 3, 3261 (1965)
50) Penczek, I., Penczek, S.: J. Polymer Sci. A-1, 8, 2465 (1970)
51) Yamashita, Y., Okada, M., Hirota, M.: Makromol. Chem. 122, 284 (1969)
52) Yamashita, Y., Ito, K., Chiba, K., Kozawa, S.: Polymer J. 3, 389 (1971)
53) Eastham, A. M.: Adv. Polymer Sci. 2, 18 (1960)
54) Goethals, E. J.: Makromol. Chem. 175, 1309 (1974)
55) Jones, G. D., MacWilliams, D. C., Braxtor, N. A.: J. Org. Chem. 30, 1994 (1965)
56) Litt, M., Levy, A., Herz, J.: J. Macromol. Sci., Chem. A 9, 703 (1975)
57) Goethals, E. J.: 4th International Symposium on Cationic Polymerization (1976)
58) Boileau, S., Champetier, G., Sigwalt, P.: Makromol. Chem. 69, 180 (1963)
59) Steiner, E. C., Pelletier, R. R., Trucks, R. O.: J. Amer. Chem. Soc. 86, 4678 (1964)
60) Wojtech, B.: Makromol. Chem. 66, 180 (1963)
61) Deibig, H., Dreiger, J., Sander, M.: Makromol. Chem. 145, 123 (1971)
62) Perret, R., Skoulios, A.: Makromol. Chem. 152, 291 (1972)
63) Taniyama, M., Nagaoka, T., Takata, T., Sayama, K.: Kogyo Kagaku Zasshi 65, 419 (1962)
64) Goodman, M., Hutchinson, J.: J. Amer. Chem. Soc. 88, 3627 (1966)
65) Lee, C. L., Johannson, O. K.: J. Polymer Sci., Polymer Chem. Ed. 14, 729, 743 (1976)
66) Mita, I., Imai, I., Kambe, H.: Makromol. Chem. 137, 143 (1970)
67) Hamiton, A., Jerome, R., Hubbert, A. J., Teyssie, Ph.: Macromolecules 6, 651 (1973)
68) Ouhadi, T., Henschen, J. M.: J. Macromol. Sci., Chem. A 9, 1183 (1975)
69) Ouhadi, T., Stevens, C., Teyssie, P.: Makromol. Chem., Suppl. 1, 191 (1975)
70) Kurata, M., Uchiyama, H., Kamada, K.: Makromol. Chem. 88, 281 (1965)
71) Croucher, T. G., Wetton, R. E.: Polymer 17, 205 (1976)
72) Pope, M. T., Weakley, T. J., Williams, R. J. P.: J. Chem. Soc. 3442, 1959
73) Jacobson, H., Stockmayer, W. H.: J. Chem. Phys. 18, 1600 (1950)
74) Flory, P. J., Semlyen, J. A.: J. Amer. Chem. Soc. 88, 3209 (1966)
75) Semlyen, J. A., Wright, P. V.: Polymer 10, 543 (1969)
76) Andrews, J. M., Semlyen, J. A.: Polymer 13, 142 (1972)
77) Andrews, J. M., Jones, F. R., Semlyen, J. A.: Polymer 15, 420 (1974)
78) Jones, F. R., Scales, L. E., Semlyen, J. A.: Polymer 15, 738 (1974)
79) Katnik, R. J., Schaefer, J.: J. Org. Chem. 33, 384 (1968)
80) Kern, R. J.: J. Org. Chem. 33, 388 (1968)
81) Dreyfuss, P., Dreyfuss, M. P.: Polymer J. 8, 811 (1976)
82) Yamashita, Y., Kitano, K., Kawakami, Y.: J. Polymer Sci., Polymer Lett. Ed., in press
83) Cooper, J., Plesch, P. H.: Chem. Comm. 1017, 1974
84) Deffieux, A., Boileau, S.: Macromolecules 9, 369 (1976)
85) Lambert, J. L., Van Ooteghem, D., Goethals, E. J.: J. Polymer Sci. A-1, 9, 3055 (1971)
86) Tsuboyama, S., Tsuboyama, K., Higashi, I., Yanagita, M.: Tetrahedron Lett. 16, 1367 (1970)
87) Goethals, E. J.: Makromol. Chem. 175, 1309 (1974)
88) Plesch, P. H., Westermann, P. H.: J. Polymer Sci. C 16, 3837 (1968)
89) Chojnowski, J., Mazurek, M., Scibiorek, M., Wilczek, L.: Makromol. Chem. 175, 3299 (1974)
90) Penczek, S., Feigin, J., Sadowska, W., Tomaszewicz, M.: Makromol. Chem. 116, 203 (1968)
91) Eastham, A. M.: Adv. Polymer Sci. 2, 18 (1960)
92) Meerwein, H., Delfs, D., Morshel, H.: Angew. Chem. 72, 927 (1960)
93) Lundberg, R. D., Kuleske, J. V., Wischemann, K. B.: J. Polymer Sci. A-1, 7, 2965 (1969)
94) Saotome, K., Kodaira, Y.: Makromol. Chem. 82, 41 (1965)
95) Deibig, H., Geiger, J., Sander, M.: Makromol. Chem. 145, 133 (1971)

40   Y. Yamashita

[96] Wilson, D. R., Beaman, R. G.: J. Polymer Sci. A-1, *8*, 2161 (1970)
[97] Gumbs, R., Penczek, S., Jagur-Grodzinski, J., Szwarc, M.: Macromolecules *2*, 77 (1969)
[98] Ito, K., Umehara, K., Yamashita, Y.: Kogyo Kagaku Zasshi *70*, 2040 (1967)
[99] Yamashita, Y., Kozawa, S., Chiba, K., Okada, M.: Makromol. Chem. *135*, 75 (1970)
[100] O'Driscoll, K. F.: J. Polymer Sci. *57*, 721 (1962)
[101] Yamashita, Y., Kozawa, S., Hirota, M., Chiba, K., Matsui, H., Hirao, A., Kodama, M., Ito, K.: Makromol. Chem. *142*, 171 (1971)
[102] Yamashita, Y., Asakura, T., Okada, M., Ito, K.: Macromolecules *2*, 613 (1969)
[103] Okada, M., Yamashita, Y., Ishii, Y.: Makromol. Chem. *80*, 196 (1964)
[104] Ito, K., Inoue, T., Yamashita, Y.: Makromol. Chem. *117*, 279 (1968)
[105] Yezrielev, A. I., Brokhina, E. L., Roskin, Ye. S.: Vysokomol. Soed. A *11*, 1670 (1969)
[106] Tidwell, R. W., Mortimer, G. A.: J. Polymer Sci. A *3*, 369 (1965)
[107] Tidwell, P. W., Mortimer, G. A.: J. Macromol. Sci.-Revs. Macromol. Chem. C *4*, 281 (1970)
[108] Joshi, R. M.: J. Macromol. Sci., Chem. A *7*, 1231 (1973)
[109] Kelen, T., Tudos, F.: J. Macromol. Sci., Chem. A *9*, 1 (1975)
[110] Kennedy, J. P., Kelen, T., Tudos, F.: J. Polymer Sci., Polymer Chem. Ed. *13*, 2277 (1975)
[111] Yamashita, Y., Asakura, T., Okada, M., Ito, K.: Makromol. Chem. *129*, 1 (1969)
[112] Minoura, Y., Mitoh, M.: Makromol. Chem. *110*, 197 (1967)
[113] Blanchard, L. P., Gabra, G. G., Malhotra, S. L.: J. Polymer Chem. Ed. *13*, 1619 (1975)
[114] Lowry, G. G.: J. Polymer Sci. *57*, 463 (1960)
[115] Wittmer, P.: Makromol. Chem. *103*, 188 (1967)
[116] Yamashita, Y., Kasahara, H., Suyama, K., Okada, M.: Makromol. Chem. *117*, 242 (1968)
[117] Kubisa, P., Penczek, S.: J. Macromol. Sci., Chem. A *7*, 1509 (1974)
[118] Durgaryan, A. A.: Vysokomol. Soedin. *8*, 790 (1966)
[119] Hazell, J. E., Ivin, K. J.: Trans. Faraday Soc. *58*, 342 (1962) *61*, 2330 (1965)
[120] Wittmer, P.: Multicomponent polymer systems. Platzer, N. A. J. (ed.). Adv. in Chem., Vol. 99, p. 140, Amer. Chem. Soc. (1970)
[121] Howell, J. A., Izu, M., O'Driscoll, K. F.: J. Polymer Sci. A-1, *8*, 699 (1970)
[122] Theil, M. H.: Macromolecules *2*, 137 (1969)
[123] Sawada, H.: J. Polymer Sci. A-1, *5*, 1383 (1967)
[124] Izu, M., O'Driscoll, K. F.: Polymer J., *1*, 27 (1970), J. Polymer Sci. A-1, *8*, 1675, 1687 (1970)
[125] Kang, K., O'Driscoll: J. Macromol. Sci. Chem. A-1, *7*, 1197 (1973)
[126] Harvey, P. E., Leonard, J.: Macromolecules *5*, 698 (1972)
[127] Bar-Zakay, S., Levy, M., Vofsi, D.: J. Polymer Sci. A-1, *5*, 965 (1967)
[128] Kagiya, T., Sumida, Y.: Bull. Chem. Soc. Japan *41*, 767, 2247 (1968)
[129] Parsiga, P. A.: J. Macromol. Sci.-Rev. Macromol. Chem. C *1*, 223 (1967)
[130] Yamshita, Y., Tsuda, T., Okada, M., Iwatsuki, S.: J. Polymer Sci. A-1, *4*, 2121 (1966)
[131] Tsuda, T., Yamashita, Y.: Makromol. Chem. *86*, 304 (1965)
[132] Okada, M., Yamashita, Y., Ishii, Y.: Makromol. Chem. *94*, 181 (1966)
[133] Plesch, P. H.: Adv. Polymer Sci. *8*, 137 (1971)
[134] Plesch, P. H., Westermann, P. H.: J. Polymer Sci. C *16*, 3837 (1968)
[135] Firat, Y., Jones, F. R., Plesch, P. H., Westermann, P. H.: Makromol. Chem. Suppl. *1*, 203 (1975)
[136] Worsfold, D. C.: J. Macromol. Sci., Chem. A *9*, 1523 (1975)
[137] Jaacks, V., Boehlke, K., Eberus, E.: Makromol. Chem. *118*, 354 (1968)
[138] Yokoyama, Y., Okada, M., Sumitomo, H.: Makromol. Chem. *176*, 795 (1975)
[139] Okada, M., Sumitomo, H., Hibino, Y.: Polymer J. *6*, 256 (1974), *7*, 511 (1975)
[140] Yamashita, Y., Okada, M., Hirota, M.: Makromol. Chem. *122*, 284 (1969)
[141] Hoehr, L., Cherdron, H., Kern, W.: Makromol. Chem. *52*, 59 (1962)
[142] Higashimura, T., Tanaka, A., Miki, T., Okamura, S.: J. Polymer Sci. A *1*, 5, 1927 (1967)
[143] Okada, M., Ikai, S., Chiba, K., Hirota, M., Yamashita, Y.: Polymer J. *1*, 1 (1970)
[144] Okada, M., Yamashita, Y.: Makromol. Chem. *126*, 266 (1969)
[145] Aoki, S., Hirota, Y., Otsu, T., Imoto, M.: Bull. Chem. Soc. Japan *38*, 1922 (1965)
[146] Yamashita, Y., Uchikawa, A.: Kogyo Kagaku Zasshi *71*, 758 (1968)

[147] Yamashita, Y., Inoue, T., Ito, K.: Makromol. Chem. *138*, 305 (1970)
[148] Yamashita, Y., Kondo, S., Ito, K.: Polymer J. *1*, 327 (1973)
[149] Tanaka, Y.: J. Macromol. Sci., Chem. A *1*, 1059 (1967)
[150] Aoki, S., Hirota, Y., Tanaka, Y., Mandai, H., Otsu, T.: J. Polymer Sci. A-1, *6*, 2585 (1968)
[151] Saegusa, T., Fujii, H., Kobayashi, S., Ando, K., Kawase, R.: Macromolecules *6*, 26, 657 (1973)
[152] Hammond, J. M., Hooper, J. F., Robertson, W. G.: J. Polymer Sci. *9*, 281 (1971)
[153] Entelis, S. G., Korovina, G. V.: Makromol. Chem. *175*, 1253 (1974)
[154] Weissermel, K., Fisher, E., Gutweiler, K., Herman, H. D.: Kunststoffe *54*, 410 (1964)
[155] Price, M. B., McAndrew, F. B.: J. Macromol. Sci. A *1*, 231 (1967)
[156] Chen, C. S. H.: J. Polymer Sci., Polymer Chem. Ed. *13*, 1183 (1975)
[157] Weissermel, K., Fischer, E., Hafner, K., Cherdron, H.: Angew. Makromol. Chem. *415*, 168 (1968)
[158] Chen, C. S. H., Edward, A. D.: J. Macromol. Sci., Chem. A *4*, 349 (1970)
[159] Bung, K. H., Fisher, E., Weissermel, K.: Makromol. Chem. *103*, 268 (1967)
[160] Kucera, M., Fichler, J.: Polymer *5*, 371 (1964)
[161] Jaacks, V.: Makromol. Chem. *101*, 33 (1967)
[162] Cherdron, H.: J. Macromol. Sci., Chem. A *6*, 1077 (1973)
[163] Chen, C. S. H., Wenger, F.: J. Polymer Sci. A *1*, 9, 33 (1971)
[164] Burg, K., Schlaf, H., Cherdron, H.: Makromol. Chem. *145*, 247 (1971)
[165] Miki, T., Higashimura, T., Okamura, S.: J. Polymer Sci. A *1*, 6, 3031 (1968), B *5*, 583 (1967)
[166] Furukawa, J.: Polymer *3*, 487 (1962)
[167] Tsuda, T., Nomura, T., Yamashita, Y.: Makromol. Chem. *86*, 301 (1965)
[168] Ito, K., Inoue, T., Yamashita, Y.: Makromol. Chem. *139*, 153 (1970)
[169] Fischer, R. P.: J. Polymer Sci. *44*, 155 (1960)
[170] Habermeier, J., Reichert, K. H., Hamann, K.: J. Polymer Sci. C *16*, 2131 (1967)
[171] Hilt, A., Reichert, K. H., Hamann, K.: Makromol. Chem. *101*, 246 (1967)
[172] Ochsner, W., Reichert, K. H.: Makromol. Chem. *150*, 1 (1971)
[173] Ito, K., Arai, K., Yamashita, Y.: Kogyo Kagaku Zasshi *74*, 117 (1971)
[174] Matyjaszewski, K., Buyle, A. M., Penczek, S.: J. Polymer Sci., Polymer Lett. Ed. *14*, 125 (1976)
[175] Tada, K., Saegusa, T., Furukawa, J.: Kogyo Kagaku Zasshi *66*, 996, 1501 (1964)
[176] Ishigaki, A., Shono, T., Hachihama, Y.: Makromol. Chem. *79*, 170 (1964)
[177] Kern, R. J., Schaefer, J.: J. Amer. Chem. Soc. *89*, 6 (1967)
[178] Borkovec, A. B.: J. Org. Chem. *23*, 828 (1958)
[179] Meerwein, H., Battenberg, E., Gold, H., Pfeil, E., Willfang, G.: J. Pr. Chem. *154*, 83 (1939)
[180] Schaeffer, J.: Macromolecules *1*, 111 (1968)
[181] Schaeffer, J., Katnik, R. J., Kern, R. J.: Macromolecules *1*, 101 (1968)
[182] Tada, K., Saegusa, T., Furukawa, J.: Makromol. Chem. *71*, 71 (1964)
[183] Yamashita, Y., Asakura, T., Okada, M., Ito, K.: Macromolecules *2*, 613 (1969)
[184] Wiemers, N., Wegner, G.: Makromol. Chem. *175*, 2719, 2743 (1974)
[185] Ito, K., Inoue, T., Yamashita, Y.: Makromol. Chem. *139*, 153 (1970)
[186] Tsuda, T., Yamashita, Y.: Makromol. Chem. *99*, 297 (1966)
[187] Saegusa, T., Ueshima, T., Imai, H., Furukawa, J.: Makromol. Chem. *79*, 22 (1964)
[188] Tada, K., Numata, Y., Saegusa, T., Furukawa, J.: Makromol. Chem. *77*, 220 (1964)
[189] Saegusa, T., Ueshima, T., Imai, H., Furukawa, J.: Makromol. Chem. *79*, 221 (1966)
[190] Yasuda, H., Tani, H.: Macromolecules *6*, 17 (1973)
[191] Araki, T., Aoyagi, T., Ueyama, N., Aoyama, T., Tani, H.: J. Polymer Sci., Polymer Chem. Ed. *11*, 699 (1973)
[192] Oguni, N., Tani, H.: J. Polymer Sci., Polymer Chem. Ed. *11*, 573 (1973)
[193] Ueyama, N., Araki, T., Tani, H.: Macromolecules *7*, 153 (1974)
[194] Saegusa, T., Matsumoto, S., Hashimoto, Y.: Polymer J. *1*, 31 (1970)
[195] Kagiya, T., Matsuda, T., Nakato, M., Hirata, R.: J. Macromol. Sci., Chem. A *6*, 1631 (1972)
[196] Saegusa, T., Ikeda, H., Fujii, H.: Polymer J. *1*, 87 (1973)

197) Saegusa, T., Kobayashi, S., Nagura, Y.: Macromolecules 7, 713 (1974)
198) Shimosaka, Y., Tsuruta, T., Furukawa, J.: Kogyo Kagaku Zasshi 66, 1498 (1963)
199) Hallensleben, M. L.: Makromol. Chem. 175, 3315 (1974)
200) Mateo, J. L., Sastre, R.: Makromol. Chem. 157, 141 (1972)
201) Hashimoto, K., Sumitomo, H.: J. Polymer Sci. A-1, 9, 107, 1189, 1747 (1971)
202) Hashimoto, K., Sumitomo, H.: Polymer J. 1, 190 (1974)
203) Odian, G., Hiraoka, L. S.: J. Makromol. Sci., Chem., A 6, 109 (1972)
204) Raes, M. C., Karabinos, J. V., Dietrich, H. J.: J. Polymer Sci. A-1, 6, 1067 (1968)
205) Yamashita, Y., Nunomoto, S., Miura, S.: Kogyo Kagaku Zasshi 69, 317 (1966)
206) Belnovskaya, G., Tchernova, Th. Dolgoplosk, B.: Europ. Polymer J. 8, 35 (1972)
207) Whipple, E. B., Green, P. J.: Macromolecules 6, 38 (1973)
208) Corno, C., Roggers, A., Salvatori, T.: Europ. Polymer J. 10, 525 (1974)
209) Corno, C., Roggers, A.: Europ. Polymer J. 12, 159 (1976)
210) Kobayashi, I., Matsuda, K.: J. Polymer Sci. A 1, 111 (1963)
211) Schirawski, G.: Makromol. Chem. 161, 69 (1972)
212) Kubanek, V., Kralicek, J., Kondelikova, J.: Angew. Makromol. Chem. 39, 77 (1974)
213) Kubanek, V.: Angew. Makromol. Chem. 46, 95 (1975)
214) Bostick, E. E.: Ring-opening polymerization. Frisch, K. C., Reegan, S. L.(ed.). New York:
      Marcel Dekker 1969, p. 342
215) Tsuruta, T.: Macromolecular Reviews 6, 179 (1972)
216) Tsuruta, T., Inoue, S., Tsukuma, I.: Makromol. Chem. 84, 298 (1965)
217) Furukawa, J., Akutsu, S., Saegusa, T.: Makromol. Chem. 94, 68 (1966)
218) Tsuruta, T., Inoue, S., Yoshida, N., Furukawa, J.: Makromol. Chem. 55, 230 (1962), 79,
      34 (1964)
219) Tsuruta, T.: Stereochemistry in macromolecules. Ketley, A. D. (ed.). New York: Marcel
      Dekker 1967, Vol. 2, p. 177
220) Furukawa, J., Kamata, Y., Yamada, K., Fueno, T.: J. Polymer Sci. C 23, 711 (1968)
221) Sepulchre, M., Spassky, N., Sigwalt, P.: Macromolecules 5, 92 (1972), Makromol. Chem.
      175, 339 (1974)
222) Dumas, P., Spassky, N., Sigwalt, P.: J. Polymer Sci., Polym. Chem. Ed. 12, 1001 (1974)
223) Oya, M., Uno, K., Iwakura, Y.: J. Polymer Sci. A-1, 10, 613 (1972)
224) Shalitin, Y., Katchalski, E.: J. Amer. Chem. Soc. 82, 1630 (1960)
225) Matsuura, K., Inoue, S., Tsuruta, Y.: Makromol. Chem. 85, 284 (1965)
226) Tsuruta, T., Inoue, S., Matsuura, K.: Biopolymers 5, 313 (1967)
227) Inoue, S., Matsuura, K., Tsuruta, T.: J. Polymer Sci. C 23, 721 (1968)
228) Iwakura, Y., Uno, K., Oya, M.: J. Polymer Sci. A-1, 6, 2165 (1968)
229) Suzuoki, K., Miyake, A., Nagoya, I., Ohizumi, C., Yamaguchi, M.: Kobunshi Ronbunshu
      31, 693, 701 (1974)
230) Suzuoki, K., Miyake, A., Nagoya, I., Ohizumi, C., Yamaguchi, M., Takeda, J.: Kobunshi
      Ronbunshu 31, 708 (1974)
231) Suzuoki, K., Miyake, A., Nagoya, I., Ohizumi, C., Yamaguchi, M., Takeda, J.: Kobunshi
      Ronbunshu 31, 715 (1974)
232) Tsuruta, T., Inoue, S., Matsuura, K.: Makromol. Chem. 63, 219 (1963)
233) Makino, T., Inoue, S., Tsuruta, T.: Makromol. Chem. 150, 137 (1971)
234) Yamashita, S., Waki, K., Yamawaki, N., Tani, H.: Macromolecules 7, 406, 410, 724 (1974)
235) Matsuura, K., Inoue, S., Tsuruta, T.: Makromol. Chem. 80, 149 (1964)
236) Makino, T., Inoue, S., Tsuruta, T.: Makromol. Chem. 150, 137 (1971)
237) Suzuoki, K., Miyake, A., Nagoya, I., Ohizumi, C., Yamaguchi, M.: Kobunshi Ronbunshu
      31, 727 (1974)
238) Kagiya, T., Narisawa, S., Ichida, T., Fukui, K., Yokota, H., Kondo, M.: J. Polymer Sci. A 1,
      4, 293 (1966)
239) Kagiya, T., Narisawa, S., Ichida, T., Fukui, K., Yokota, H.: J. Polymer Sci. A 1, 4, 2171
      (1966), 5, 1645, 2031 (1967)
240) Furukawa, J., Iseda, Y., Saegusa, T., Fujii, H.: Makromol. Chem. 89, 263 (1965)
241) Inoue, S., Koinuma, H., Tsurata, T.: J. Polymer Sci. B 7, 287 (1969)

242) Inoue, S., Koinuma, H., Tsuruta, T.: Makromol. Chem. *130*, 212 (1969)
243) Kobayashi, M., Inoue, S., Tsuruta, T.: Macromolecules *4*, 658 (1971), Makromol. Chem. *169*, 69 (1973), J. Polymer Sci., Polymer Chem. Ed. *11*, 2383 (1973)
244) Kuran, W., Pasynkiewicz, S., Skupinska, J.: Makromol. Chem. *177*, 11, 1283 (1976)
245) Inoue, S.: Chemtech. *1976*, 588
246) Kagiya, T., Matsuda, T.: Polymer J. *3*, 398 (1971)
247) Soga, K., Ikeda, S.: J. Polymer Sci., Polymer Lett. Ed. *11*, 479 (1973)
248) Soga, K., Ikeda, S.: J. Polymer Sci., Polymer Chem. Ed. *12*, 121 (1974)
249) Soga, K., Hosoda, S., Ikeda, S.: Makromol. Chem. *175*, 3309 (1974)
250) Saegusa, T., Kobayashi, S., Kimura, Y.: Macromolecules *10*, 68 (1977)
251) Yamazaki, N., Higashi, F., Iguchi, T.: J. Polymer Sci., Polymer Lett. Ed. *12*, 517 (1974)
252) Soga, K., Sato, M., Imaura, H., Ikeda, S.: J. Polymer Sci., Polymer Lett. Ed. *13*, 167 (1975), Polymer Chem. Ed. *14*, 677 (1976)
253) Schaeffer, J., Kern, R. J., Katnick, R. J.: Macromolecules *1*, 107 (1968)
254) Aso, C., Tagami, S., Kunitake, T.: Kobunshi Kagaku *23*,
255) Yamashita, Y., Uchikawa, A., Yoshida, K., Kobayashi, T.: Makromol. Chem. *105*, 292 (1967)
256) Hilt, A., Reichert, K. H., Hamann, K.: Makromol. Chem. *101*, 246 (1967)
257) Hsieh, H. L.: J. Macromol. Sci., Chem. A *7*, 1525 (1973)
258) Natta, G., Mazzanti, G., Pregaglia: J. Polymer Sci. *58*, 1021 (1962)
259) Miller, R. G. J., Nield, E., Turner-Jones, A.: Chem. Ind. *27*, 181 (1962)
260) Natta, G., Mazzanti, G., Pregaglia, G. F., Binaghi, M.: Makromol. Chem. *44/46*, 537 (1961)
261) Fischer, R. F.: J. Polymer Sci. *44*, 155 (1960), Ind. Eng. Chem. *52*, 321 (1960)
262) Schwenk, E., Gulbins, K., Roth, M., Benzing, G., Maysenholder, R., Hamann, K.: Makromol. Chem. *51*, 53 (1962)
263) Kern, R. J.: J. Amer. Chem. Soc. *89*, 6 (1967)
264) Tada, K., Saegusa, T., Furukawa, J.: Kogyo Kagaku Zasshi *68*, 1985 (1965)
265) Saegusa, T.: Chemtech. 295, 1975
266) Saegusa, T., Kobayashi, S., Kimura, Y., Ikeda, H.: J. Macromol. Sci., Chem. A *9*, 641 (1975)
267) Saegusa, T., Kobayashi, S., Kimura, Y.: Macromolecules *7*, 1 (1974)
268) Saegusa, T., Kobayashi, S., Kimura, Y.: Macromolecules *7*, 139 (1974)
269) Saegusa, T., Kimura, Y., Sano, K., Kobayashi, S.: Macromolecules *7*, 546 (1974)
270) Saegusa, T., Kimura, Y., Sawada, S., Kobayashi, S.: Macromolecules *7*, 956 (1974)
271) Saegusa, T., Kobayashi, S., Kimura, Y.: Macromolecules *8*, 374 (1975), *9*, 724 (1976), *10*, 64 (1977)
272) Saegusa, T., Ikeda, H., Hirayanagi, S., Kimura, Y., Kobayashi, S.: Macromolecules *8*, 259 (1975)
273) Saegusa, T., Kobayashi, S., Furukawa, J.: Macromolecules *8*, 703 (1975), *9*, 728 (1976), *10*, 73 (1977)
274) Kagiya, T., Narisawa, S., Manabe, K., Fukui, K.: J. Polymer Sci. B *3*, 617 (1965)
275) Kagiya, T., Narisawa, S., Manabe, K., Konata, M., Fukui, K.: J. Polymer Sci. A 1, *4*, 208 (1966)
     Grundschober, F., Sambeth, J.: J. Polymer Sci. C *16*, 2087 (1967)
276) Hashimoto, S., Yamashita, T., Ono, M.: Kobunshi Ronbunshu *33*, 373 (1976)
277) Bailey, W. J., Okamoto, Y.: Polymer Preprints *12*, 177 (1971)
278) Bailey, W. J., Shah, K.: Polymer Preprints *15*, 587 (1974)
279) Goodman, M., Ingwall, R. T., Pierre, S. St.: Macromolecules *9*, 1 (1976)
280) Katakai, R., Oya, M., Toda, F., Uno, K., Iwakura, Y.: Macromolecules *6*, 827 (1973)
281) Allport, D. C., Janes, W. H.: Block copolymers. London: Applied Science Publishers 1973
282) Ceresa, R. J.: Block and Graft Copolymerization, Vol. 1, 2. New York: Wiley Interscience 1973, 1976
283) Henderson, J. F., Szwarc, M.: Macromolecular Reviews *3*, 317 (1968), New York: Interscience
284) Hayashi, K., Marvel, C. S.: J. Polymer Sci. A *2*, 2571 (1964)

285) Morton, M., Fetters, L. J.: Macromolecular Reviews 2, 71 (1967)

286) Wentz, C. A., Hopper, E. E.: Ind. Eng. Chem., Prod. Res. Dev. 6, 209 (1967)

287) Reed, S. F., Jr.: J. Polymer Sci. A-1, 10, 1187 (1972)

288) Yamashita, Y., Hirota, M., Matsui, H., Hirata, A., Nobutoki, K.: Polymer J. 2, 43 (1971)

289) Franta, E., Reibel, L., Lehmann, J., Penczek, S., Dobrogozcz, W.: 4th International Symposium on Cationic Polymerization, Akron (1976)

290) Smith, S., Hubin, A. J.: J. Macromol Sci., Chem. A 7, 1399 (1973)

291) Noll, W.: Chemistry and technology of silicones. Academic Press 1968

292) Furukawa, J., Yamashita, S., Ido, S.: Polymer chemistry of synthetic elastomers. Kennedy, J. P., Tornqvist, E. (ed.). New York: Interscience 1968, Vol. 11, p. 843

293) Reed, S. F.: J. Polymer Sci. A-1, 9, 2029, 2147 (1971)

294) Battaerd, H. A., Tregear, G. W.: Graft copolymers. New York: Interscience 1967

295) Richaeds, D. H., Szwrc, M.: Trans. Faraday Soc. 55, 1644 (1959)

296) Skoulios, A., Finaz, G., Parrod, J.: Compt. Rend. 251, 739 (1960)

297) Baer, M.: J. Polymer Sci. A 2, 417 (1964)

298) O'Malley, J. J., Crystal, R. G., Erhardt, P. F.: Polymer Preprints 11, 796 (1969)

299) O'Malley, J. J., Marchessault, R. H.: Macromolecular Syntheses, 4, 35 (1972)

300) Wesslen, B., Mansson, P.: J. Polymer Sci., Polymer Chem., Ed. 13, 2545 (1975)

301) Hirata, E., Ijitsu, T., Soen, T., Hashimoto, T., Kawai, H.: Polymer 16, 249 (1975)

302) Seow, P. K., Gallot, Y., Skoulios, A.: Makromol. Chem. 176, 3153 (1975)

303) Kikuchi, M., Kakurai, T.: Kobunshi Ronbunshu 31, 743 (1974)

304) Tomoi, M., Shibayama, Y., Kakiuchi: Polymer J. 8, 190 (1976)

305) Furukawa, J., Saegusa, T., Mise, N.: Makromol. Chem. 38, 244 (1960);
Galin, J. C.: Makromol. Chem. 124, 118 (1969)

306) Suzuki, T., Murakami, Y., Tsuji, Y., Takegami, Y.: J. Polymer Sci., Polymer Lett. Ed. 14, 675 (1976)

307) Minoura, Y., Nakano, A.: Macromolecular Syntheses 4, 25 (1972)

308) Tobolsky, A. V., Rembaum, A.: J. Appl. Polymer Sci. 8, 301 (1964)

309) Galin, J., Galin, M., Calme, P.: Makromol. Chem. 134, 273 (1970)

310) Furukawa, J., Takamori, S., Yamashita, S.: Angew. Makromol. Chem. 1, 92 (1967)

311) Laverty, J. J., Gardlaud, A. G.: Polymer Preprints 15, (2), 306 (1974)

312) Tahan, M., Zilkha, A.: Eur. Polymer J. 5, 347 (1969)

313) Erza, G., Zilkha, A.: J. Polymer Sci. A 1, 8, 1343 (1970)

314) Erza, F., Zilkha, A.: Europ. Polymer J. 6, 403, 1305 (1970)

315) Tahan, M., Zilkha, A.: J. Polymer Sci. A-1, 7, 1815, 1825 (1969)

316) Lundsted, L. G., Schmolka, I. R.: Block and graft copolymerization. Ceresa, R. J. (ed.). London: John Wiley 1976, Vol. 2, p 1

317) Saegusa, T., Matsumoto, S., Hashimoto, Y.: Macromolecules 3, 377 (1970)

318) Yamashita, Y., Chiba, K.: Polymer J. 4, 200 (1973)

319) Yamashita, Y., Okada, M., Hirota, M.: Angew, Makromol. Chem. 9, 136 (1969)

320) Okada, M., Sumitomo, H., Kakezawa, T.: Makromol. Chem. 162, 285 (1972)

321) Aoki, S., Otsu, T., Imoto, M.: Kogyo Kagaku Zasshi 67, 971, 1955 (1964)

322) Aoki, S., Otsu, T., Haruta, Y., Imoto, M.: Kogyo Kagaku Zasshi 67, 1953 (1964)

323) Franta, E., Reibel, L., Lehmann, J., Penczek, S., Dobrogozcy, W.: 4th International Symposium on Cationic Polymerization, Akron (1976)

324) Franta, E., Afshar-Taromi, F., Rempp, P.: Makromol. Chem. 177, 2191 (1976)

325) Dreyfuss, P., Kennedy, J. P.: 4th International Symposium on Cationic Polymerization, Akron (1976)

326) Dreyfuss, P., Kennedy, J. P.: J. Polymer Sci., Polymer Lett. Ed. 14, 135, 139 (1976)

327) Noro, K., Kawazura, H., Moriyama, T., Yoshioka, S.: Makromol. Chem. 83, 35 (1965)

328) Smith, W. E., Galiano, F. R., Rankin, D., Mantell, G. J.: J. Appl. Polymer Sci. 10, 1659 (1966)

329) Takida, H., Noro, K.: Kobunshi Kagaku 21, 459, 467, 724 (1964)

330) Mita, I., Imai, I., Kambe, H.: Makromol. Chem. 137, 143 (1970)

331) Hashimoto, K., Sumitomo, H.: J. Polymer Sci. A 1, 7, 1549 (1969)

332) Godfrey, R. A., Miller, G. W.: J. Polymer Sci. A 1, 7, 2387 (1969)
333) Shiota, T., Hayashi, K.: J. Appl. Polymer Sci. 11, 753, 773, 791 (1967)
334) Shiota, T., Hayashi, K.: J. Appl. Polymer Sci. 12, 2421, 2441, 2463 (1968)
335) Shiota, T., Hayashi, K.: J. Appl. Polymer Sci. 13, 1749, 2447 (1969)
336) Yamashita, Y., Hane, T.: J. Polymer Sci., Polymer Chem. 11, 425 (1973)
337) Yamashita, Y., Nakamura, Y., Kojima, S.: J. Polymer Sci., Polymer Chem. 11, 823 (1973)
338) Sundet, S. A., Thamm, R. C., Meyer, J. M., Buck, W. H., Caywood, S. W., Subramanian, P. M., Anderson, B. C.: Macromolecules 9, 371 (1976)
339) Foss, R. P., Jacobson, H. W., Cripps, H. N., Sharkey, W. H.: Macromolecules 9, 373 (1976)
340) Saotome, K., Kodaira, Y.: Makromol. Chem. 82, 41 (1965)
341) Pilato, L. A., Koleske, J. V., Joesten, B. L., Robeson, L. M.: Polymer Preprints 17, (2), 824 (1976)
342) Nobutoki, K., Sumitomo, H.: Bull. Chem. Soc., Japan 40, 1741 (1967)
343) Tabuchi, T., Nobutoki, K., Sumitomo, H.: Kogyo Kagaku Zasshi 71, 1926 (1968)
344) Perret, R., Skoulios, A.: Makromol. Chem. 156, 143 (1972)
345) Hsieh, H. L.: Polymer Preprints 17, 200 (1976)
346) Nevin, R. S., Pearce, E. M.: J. Polymer Sci. B 3, 487 (1965)
347) Boileau, S., Sigwalt, P.: Compt. Rend. 261, 132 (1965)
348) Morton, M., Kammereck, R. F., Fetters, L. J.: Macromolecules 4, 11 (1971), Brit. Polymer J. 3, 120 (1971)
349) Gourdenne, A., Sigwalt, P.: Europ. Polymer J. 3, 481 (1967)
350) Gourdenne, A., Sigwalt, P.: Bull. Soc. Chem., France 10, 3685 (1967)
351) Gourdenne, A.: Polymer Preprints 12, (2), 129 (1971)
352) Gourdenne, A., Boileau, S., Fontanille, M., Sigwalt, P.: Polymer Preprints 10, 826 (1969), Makromol. Chem. 131, 7 (1970)
353) Morton, M., Kammereck, R. F.: J. Amer. Chem. Soc. 92, 3217 (1970)
354) Oswald, A. A.: Polymer Preprints 13, 57 (1972)
355) Mackillop, D. A.: J. Polymer Sci. B 8, 199 (1970)
356) Morton, M., Mikesell, S. L.: J. Macromol. Sci., Chem. A 7, 1391 (1973)
357) Cooper, W., Hale, P., Walker, J. S.: Polymer 15, 175 (1974)
358) Hale, P., Pope, G. A.: Europ. Polymer J. 11, 677 (1975)
359) Boileau, S., Sigwalt, P.: Makromol. Chem. 171, 11 (1973)
360) Lambert, J. L., Goethals, E. J.: Makromol. Chem. 133, 289 (1970)
361) Yamashita, Y., Matsui, H., Ito, K.: J. Polymer Sci., Polymer Chem. Ed. 10, 3577 (1972)
362) Yamashita, Y., Murase, Y., Ito, K.: J. Polymer Sci., Polymer Chem. Ed. 11, 435 (1973)
363) Hergenrother, W. L., Ambrose, R. J.: J. Polymer Sci., Polymer Chem. Ed. 12, 2613 (1974)
364) Matzner, M., Schober, D. L., McGrath, J. E.: Europ. Polymer J., 9, 469 (1973)
365) Shalabt, S. W., Pearce, E. M., Reimschussel, H. K.: Ind, Eng. Chem. Res. Develop. 12, 128 (1973)
366) Ide, F., Haseyama, A.: J. Appl. Polymer Sci. 18, 963 (1974)
367) Litt, M., Herz, J., Turi, E.: Polymer Preprints 10, (2), 905 (1969)
368) Saegusa, T., Ikeda, H.: Macromolecules 6, 805 (1973)
369) Saegusa, T., Ikeda, H., Fujii, H.: Macromolecules 6, 315 (1973)
370) Saegusa, T., Kobayashi, S., Yamada, A.: Macromolecules 8, 390 (1975)
371) Goodman, M., Hutchison, J.: J. Amer. Chem., Soc. 88, 3627 (1966)
372) Oya, M., Katagai, R., Uno, K., Iwakura, Y.: Kogyo Kagaku Zasshi 73, 2371 (1970)
373) Iwakura, Y., Uno, K., Oya, M.: J. Polymer Sci. A 1, 6, 2165 (1968)
374) Oya, M., Uno, K., Iwakura, Y.: Bull. Chem. Soc. Japan 43, 1788 (1970)
375) Yamashita, Y., Iwaya, Y., Ito, K.: Makromol. Chem. 176, 1207 (1975)
376) Billot, J., Douy, A., Gallot, B.: Makromol. Chem. 177, 1889 (1976)
377) Perly, B., Douy, A., Gallot, B.: Makromol. Chem. 177, 2569 (1976)
378) Mori, S., Iwatsuki, M.: Kobunshi Kagaku 30, 39 (1973)
379) Seeney, C. E., Harwood, H. J.: J. Macromol. Sci., Chem. A 9, 779 (1975)
380) Reibel, L., Spach, G.: Bull. Soc. Chim. France 1972, 1025
381) Morton, M., Rembaum, A., Bostick, E. E.: J. Appl. Polymer Sci. 8, 2707 (1964)

382) Minoura, Y., Mitoh, M., Tabuse, Y., Yamada, Y.: J. Polymer Sci. A 1, 7, 2753 (1969)
383) Lee, C. L., Johannson, O. K.: J. Polymer Sci. A 1, 4, 3013 (1966)
384) Saam, J. C., Gordon, D. J., Lindsey, S.: Macromolecules 3, 1 (1970)
385) Owen, M. J., Kendrick, T. C.: Macromolecules 3, 458 (1970)
386) Davies, W. G., Jones, D. P.: Polymer Preprints 11, 447 (1970)
387) Dean, J. W.: J. Polymer Sci. B 8, 677 (1970)
388) Zilliox, J. G., Roovers, J. E., Bywater, S.: Macromolecules 8, 573 (1975)
     Marsiat, A., Gallot, Y.: Makromol. Chem. 176, 1641 (1975)
389) Gaines, G. L. Jr., Bender, G. W.: Macromolecules 5, 82 (1972)
390) Jones, F. R.: Europ. Polymer J. 10, 249 (1974)
391) Bostick, E. E.: Polymer Preprints 10, (2), 877 (1969)
392) Riches, K., Haward, R. N.: Polymer 9, 103 (1968)
393) Bamford, C. H., Jenkins, A. D., Wayne, R. P.: Trans. Faraday Soc. 56, 932 (1956)
394) Berger, G., Levy, M., Vofsi, D.: J. Polymer Sci. B 4, 183 (1966)
395) Yamashita, Y., Hirota, M., Nobutoki, K., Nakamura, Y., Hirao, A., Kozawa, S., Chiba, K.,
     Matsui, H., Hattori, G., Okada, M.: J. Polymer Sci. B 8, 483 (1970)
396) Takahashi, A., Yamashita, Y.: Polymer Preprints 15, (1), 184 (1974)
397) Yamashita, Y., Nobutoki, K., Nakamura, Y., Hirota, M.: Macromolecules 4, 548 (1971)
398) Greber, G., Reese, E., Balciunas, A.: Farbe u. Lack 70, 249 (1964)
399) Morton, M., Kesten, Y., Fetters, L. J.: Polymer Preprints 15, (2), 175 (1974)
400) Greber, G., Balciunas, A.: Makromol. Chem. 79, 149 (1964)
     Vaughn, H. A., Kambour, R. P., Le Grand, D. G.: J. Polymer Sci. B 7, 569 (1969)
401) Noshay, A., Matzner, M., Williams, T. C.: Ind. Eng. Chem., Prod. Res. Dev. 12, (4), 268
     (1973)
402) Noshay, A., Matzner, M., Merriam, C. N.: J. Polymer Sci. A 1, 9, 3147 (1971)
403) Noshay, A., Matzner, M., Karoly, G., Stampa, G. B.: J. Appl. Polymer Sci. 17, 619 (1973)
404) Matzner, M., Noshay, A., Robeson, L. M., Merriam, C. N., Barclay, R., McGrath, J. E.:
     Appl. Polymer Symp. 22, 143 (1973)
405) Pacansky, T. J.: Polymer Preprints 17, (2), 564 (1976)
406) Tanquary, A. C., Burks, R. E., jr., Jackson, M. V.: J. Polymer Sci. Polymer Chem. Ed., 13,
     119 (1975)
407) Greber, G.: Makromol. Chem. 101, 104 (1967)
408) Plumb, J. B., Atherton, J. H.: Block copolymers. Allport, D. C., Janes, W. H. (ed.). Lon-
     don: Applied Science Publishers 1973 p. 305
409) Twaik, M. A., Tahan, M., Zilkha, A.: J. Polymer Sci. A-1, 7, 2469 (1969)
410) Thierry, A., Skoulios, A.: Makromol. Chem. 177, 319 (1976)
411) Ide, F., Hasegawa, A.: J. Appl. Polymer Sci. 18, 963 (1974)
412) Bi, L., Fetters, L. J.: Macromolecules 9, 732 (1976)

Received January 17, 1978
H. Fujita (editor)

Ring-Opening Polymerization of Bicyclic Acetals, Oxalactone, and Oxalactam

Hiroshi Sumitomo and Masahiko Okada

Faculty of Agriculture, Nagoya University, Chikusa-ku, Nagoya 464, Japan

Reaction mechanisms of ring-opening polymerization and structures and properties of resulting polymers of bicyclic acetals including 6,8-dioxabicyclo[3.2.1]octane (abbreviated to DBO) and 6,8-dioxabicyclo[3.2.1]oct-3-ene (DBOE), bicyclic oxalactone 6,8-dioxabicyclo[3.2.1]octan-7-one (DBOO), bicyclic oxalactam 8-oxa-6-azabicyclo[3.2.1]octan-7-one (BOL), and related hetero bicyclic compounds are reviewed. Polysaccharide analogues were prepared from DBO and DBOE. Novel macrocyclic oligoesters were obtained in high yields from DBOO. BOL yielded easily a novel hygroscopic polyamide membrane of high molecular weight.

Table of Contents

I. Introduction

Ring-opening polymerization has played an important role in systematic approaches
to the chemical synthesis of polysaccharides and the clarification of their biomedically
interesting characteristics in relation to their own molecular structures. The early

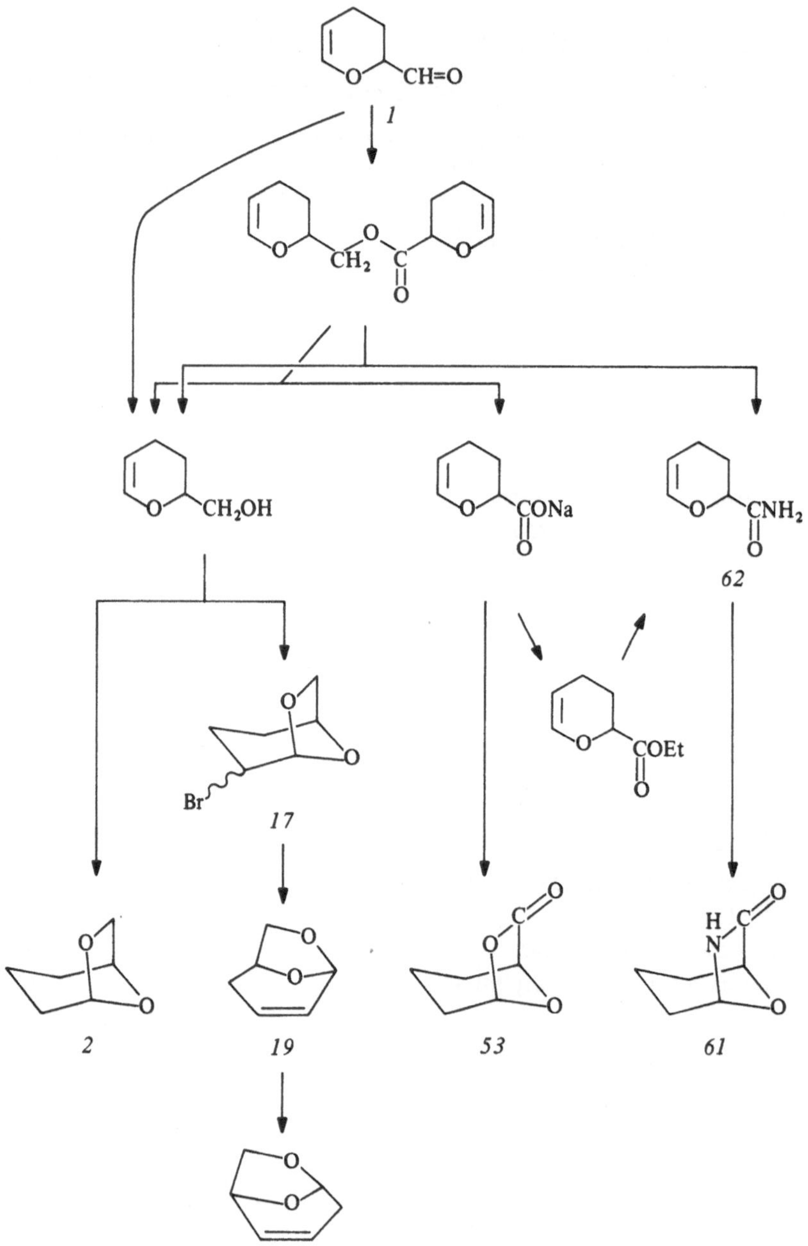

Scheme 1

reports by Korshak[1] and Schuerch[2] on the stereospecific polymerization of
1,6-anhydro-2,3,4-tri-0-methyl-D-glucopyranose and Schuerch's subsequent successful works[3, 4] on the polymerization of anhydro sugar derivatives have been worth
noting.

In connection with studies on the ring-opening polymerization of cyclic acetals,
we have undertaken investigations on the polymerization of bicyclic acetals, bicyclic
oxalactone, and bicyclic oxalactam, which yield polysaccharide analogs, macrocyclic
oligoesters, and a hydrophilic polyamide, respectively, some of which can be expected
to be useful as novel speciality polymers. The monomers employed in the studies
were prepared via synthetic routes presented in Scheme 1, starting from 3,4-dihydro-
2H-pyran-2-carbaldehyde (acrolein dimer) 1.

II. Polymerization of Bicyclic Acetals

The replacement of one or more of the hydrogens of the 1,3-dioxolane and 1,3-di-
oxepane ring by alkyl groups reduces the polymerizability to a great extent. Thus,
2-methyl- and 2,4-dimethyl-1,3-dioxolanes[5, 6], and 2,4-dimethyl-1,3-dioxepane[7]
do not polymerize at all under ordinary conditions, or at best, they give oligomeric
products in poor yield, in striking contrast to relatively high polymerizability of unsubstituted 1,3-dioxolane and 1,3-dioxepane. Such substituent effects are well consistent with the prediction based on thermodynamic consideration[8]. However, it is
expected that bicyclic acetals, which can be regarded as disubstituted cyclic acetals,
would exhibit enhanced reactivity owing to their appreciable ring strain and bond
repulsion which would be relieved by ring-opening polymerization. In contrast to
the abundance of papers dealing with ring-opening polymerization of monocyclic
acetals such as 1,3-dioxolane and 1,3-dioxepane, there have been only a small number of papers concerned with the polymerization of bicyclic acetals, although a considerable number of papers on the polymerization of anhydro sugar derivatives have
been published. In this section, however, description of the polymerization of an-
hydrosugar derivatives has been eliminated because of the restricted length of this
article. Those who are interested in this field should refer to the excellent reviews
by Schuerch[3, 4].

1. 6,8-Dioxabicyclo[3.2.1]octanes

Polymerization of 6,8-dioxabicyclo[3.2.1]octane, 2, has been most extensively
studied among bicyclic acetals. This monomer is readily prepared from 3,4-dihydro-
2H-pyran-2-carbaldehyde 1 by reduction with sodium borohydride followed by acid-

catalyzed cyclization[9]. Polymerization of *2* was investigated independently by three different groups.

Kops[10] reported that the polymerization of *2* initiated with phosphorus pentafluoride in methylene chloride gave only oligomeric materials with number average molecular weight of 400–600, and that solid state polymerization gave rise to polymers with somewhat higher molecular weight.

Hall and Steuck[11] polymerized *2* with a variety of Lewis and Brönsted acids or oxonium salts. The best conditions for the polymerization proved to be the use of phosphorus pentafluoride in methylene chloride solution at −78 °C. Yields of methanol-insoluble polymers ranging from 68 to 84% were obtained with inherent viscosities of 0.26–0.33 dl/g. Lower or higher temperatures gave lower yields. Tetrahydrofuran as solvent at −78 °C gave 68–92% yields of materials having inherent viscosities of 0.12–0.14 dl/g. No incorporation of tetrahydrofuran into the polymer occurred.

The present authors[12, 13] also studied the polymerization of *2* with boron trifluoride etherate as the initiator. High molecular weight polymers with intrinsic viscosities up to 1.87 dl/g were obtained in methylene chloride at −78 °C. The polymers melted at 160–180 °C, and showed some crystallinity as observed by X-ray diffraction.

Figure 1 shows the NMR spectra of the methanol-insoluble polymers prepared at 0 °C (A) and −78 °C (B). The peaks at δ 4.85 and 4.40 ppm in the spectrum A can be assigned, respectively, to the equatorial and axial acetal protons of two structural

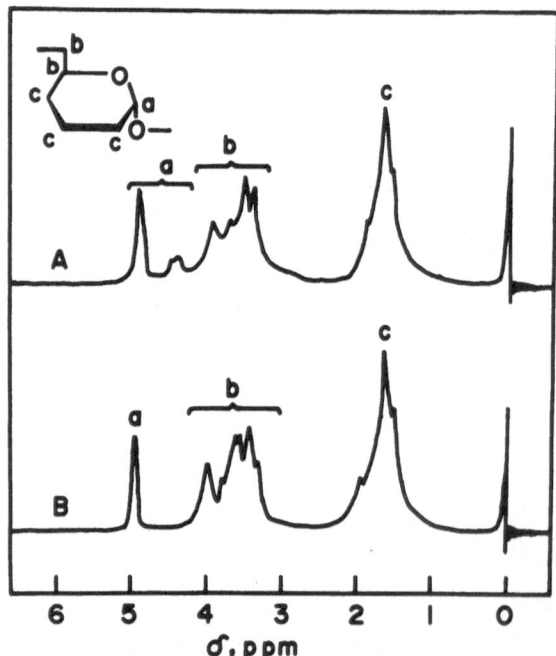

Fig. 1. ^{1}H-NMR spectra of poly-6,8-dioxabicyclo[3.2.1]octane prepared in methylene chloride at A, 0 °C and B, − 78 °C. Solvent, CDCl$_3$; concn., ca. 5%; room temp.; 100 MHz

units, *4* and *5*[14]. In spectrum B, however, the peak at δ 4.40 ppm disappears nearly completely. This means that the polymer obtained at −78 °C contains exclusively the structural unit *4* and hence it is a stereoregular polymer having the same backbone structure as that of the natural dextran *6*.

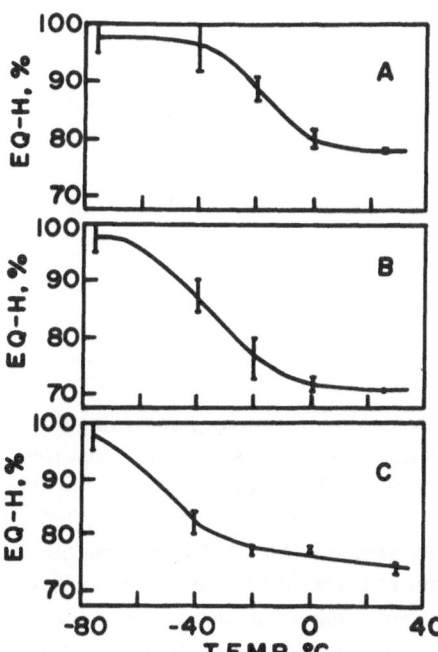

As a measure of the stereoregularity, an index EQ-H% was defined as the precent of the equatorial acetal protons to the total acetal protons. Figure 2 illustrates the temperature dependence of EQ-H%'s of the polymer obtained in toluene (A), methylene chloride (B), and 1-nitropropane (C). No significant difference is observed at

Fig. 2. Temperature dependence of EQ-H%
Solvent, A, toluene; B, methylene chloride;
C, 1-nitropropane[13]

−78 °C and at temperatures above 0 °C among the EQ-H%'s of the polymers from the three series of experiments with different solvents. However, in the temperature range from −40 to −20 °C, the EQ-H%'s of the polymers from these series decrease in the order: toluene > methylene chloride > 1-nitropropane. The EQ-H%'s of the polymers obtained at 0 °C did not depend upon the conversions, but those of the polymers prepared at −40 °C decreased slightly with increasing conversions, and when the reaction mixture was allowed to stand at $-40\,°C$ for a prolonged time after it had arrived near the equilibrium conversion, the EQ-H% of the polymer dropped to a constant value of 75–77%[13].

Hall et al.[15] estimated the conformational equilibrium for the structural units in the polymer of 2 using the numerical parameters determined for carbohydrates[16]. For a trans-1,3-tetrahydropyranoside, conformer 8 is calculated to be more stable than 7 by 9.2 kJmol^{-1} and would therefore occur almost exclusively (ca. 98%) at equilibrium. For a cis-1,3-tetrahydropyranoside unit, the anomeric effect favors conformer 9, but its severe syn-axial interaction between alkoxy and alkyl groups would highly favor 10 (ca. 99%).

7   8

$$\Delta G° = -9.2 \text{ kJmol}^{-1}$$

$$K = \frac{8}{7} = 42$$

9   10

$$\Delta G° = -9.4 \text{ kJmol}^{-1}$$

$$K = \frac{10}{9} = 81$$

The stereoregularity of the polymers provides useful information regarding the mechanism of this polymerization. As depicted in Scheme 2, 2 approaches from the opposite side of the positively charged oxygen atom to the acetal carbon atom of the terminal unit 11 and cleaves the carbon-oxygen bond of the oxonium ion by a S_N2 type reaction (broken arrow a) to regenerate the oxonium ion 12. The flipping of the tetrahydropyran ring of the penultimate unit to form the structure 13 seems less likely to take place, because of the unfavorable conformational energy of this form. The addition of 2 from the same side of the positively charged oxygen atom appears improbable because of severe steric hindrance. Therefore, the fact that the stereoregular polymers possessing acetal hydrogens in the equatorial positions are obtained at $-78\,°C$ is reasonably explained in terms of an S_N2 mechanism involving the cyclic trialkyloxonium ion[13].

There are two possible ways to interpret the decrease in the EQ-H% of the polymers with rise in temperature and/or the polarity of the solvent. The first is the concept that the growing chain end in the cationic polymerization of 2 consists of the cyclic trialkyloxonium ion 11 and the oxycarbenium ion 14, the latter of which is

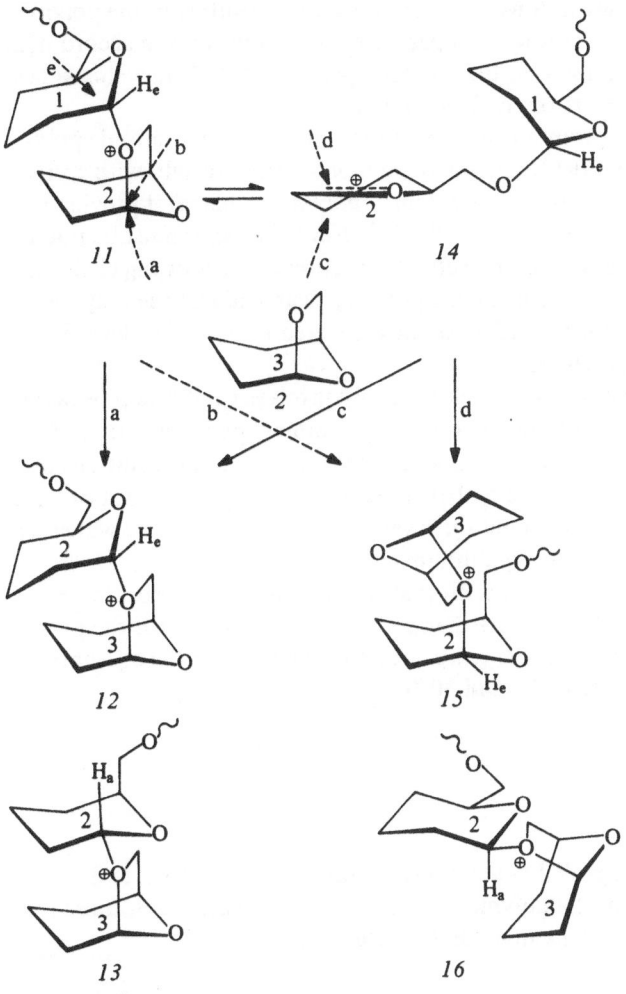

Scheme 2[13]

derived from the former by monomolecular ring-opening reaction. This concept was proposed by Zacoval and Schuerch[17] to explain the observed decrease in the specific rotation of the polymers obtained at high temperatures in the cationic polymerization of 1,6-anhydro-2,3,4-tri-0-benzyl-D-mannopyranose. The cyclic trialkyloxonium ion *11* and the oxycarbenium ion *14* are probably in equilibrium. The former, as described above, reacts stereospecifically to afford the structure *12* at one rate, and the latter nonstereospecifically to yield the structures *12* and *16* at another rate. Therefore, the results in Fig. 2 can be rationalized by assuming that the contribution of the oxy-carbenium ion *14* to the formation of the polymer becomes greater with the rise in temperature and/or the polarity of the solvent[13].

The second explanation for the formation of the stereoirregular polymer is that the propagation of *2* is essentially an S_N1 type reaction, and that attack d (broken arrow) in *14* is depressed by the interaction of the outer-ring oxygen atom with the positively charged carbon atom of the oxycarbenium ion from the upper side of the tetrahydropyran ring, especially at lower temperatures, thus resulting in the polymers with increasing stereoregularity at lower temperatures. This concept is in accord with the ^{1}H- and ^{13}C-NMR observation that the growing species of 1,3-dioxacycloalkanes in the cationic polymerization are oxycarbenium ions[18, 19].

The oxycarbenium ionic nature of the growing species of 2 in its cationic polymerization is also suggested by the finding that 2 copolymerized readily with styrene in methylene chloride at 0 °C to afford a random copolymer, whereas it provided products with a highly block character at −78 °C[20]. Nevertheless, it must be noted here that side reactions such as oxonium exchange reaction at the growing chain end of 2, or acetal exchange reaction involving the growing chain end and the polymer chain also can be responsible for the reduction of the stereoregularity of the polymer of 2 at relatively higher temperatures.

Cationic polymerization of cyclic acetals generally involves equilibrium between monomer and polymer. The equilibrium nature of the cationic polymerization of 2 was ascertained by depolymerization experiments: Methylene chloride solutions of the polymer ($[P]_0$ = 1.76 and 1.71 base-mol/l) containing a catalytic amount of boron trifluoride etherate were allowed to stand for several days at 0 °C to give 2 which was in equilibrium with its polymer. The equilibrium concentrations ($[M]_e$ = 0.47 and 0.46 mol/l) were in excellent agreement with that found in the polymerization experiments under the same conditions. The thermodynamic parameters for the polymerization of 1 were evaluated from the temperature dependence of the equilibrium monomer concentrations between −20 and 30 °C.

$$\Delta H_{ss} = -17.5 \pm 1.7 \text{ kJmol}^{-1}$$
$$\Delta S_{ss}^{\circ} = -59.4 \pm 5.9 \text{ Jmol}^{-1}\text{K}^{-1}$$

The monomer 2 can be regarded as a 2,4-disubstituted-1,3-dioxolane, and the thermodynamic parameters for the polymerization may be compared with those for the polymerization of 1,3-dioxolane in methylene chloride[21].

$$\Delta H_{ss} = -21.3 \pm 0.8 \text{ kJmol}^{-1}$$
$$\Delta S_{ss}^{\circ} = -79.8 \pm 5.0 \text{ Jmol}^{-1}\text{K}^{-1}$$

The ΔH_{ss} value for 2 is less negative than that for 1,3-dioxolane, in accordance with the predictions from the thermodynamic consideration for the hypothetical polymerizations of cycloalkanes[8] and also from rotational isomer models for the polymerization of alkyl-substituted 1,3-dioxolanes[22]. Enthalpy change for the polymerization of cyclic monomers, in general, reflects the strain involved in the monomer. The relatively small negative value of ΔH_{ss} for 2 implies that the monomer 2, although it is bicyclic, contains a relatively small strain. In this regard, Hall[23] conjectured from molecular model inspection and polymerization experiments that monomers having a bicyclo[3.2.1] skeleton were almost free from angular distortion and non-bonded repulsion. Entropy change for the polymerization of 2 is less negative than that of 1,3-dioxolane, thus favoring the polymerization of the former. This is opposed to the prediction from thermodynamical consideration of cycloalkanes that the introduction of a methyl substitutent to a cycloalkane ring makes the entropy change for the polymerization more negative[8].

1-Methyl- and 5,7-dimethyl-6,8-dioxabicyclo[3.2.1]octanes were polymerized in bulk in the presence of a large amount of boron trifluoride[24]. The resulting polymers having molecular weights of 900–1130 melted at 95–97 and 102–105 °C, respectively.

17a   *17b*

Polymerization of 4-bromo-6,8-dioxabicyclo[3.2.1]octane *17* in dichloromethane solution at −78 °C with phosphorus pentafluoride as initiator gave a 60% yield of polymer having an inherent viscosity of 0.10 dl/g[11]. Although it is not described explicitly, the monomer used seems to be a mixture of the stereoisomers, *17a* and *17b*, in which the bromine atom is oriented trans and cis, respectively, to the five-membered ring of the bicyclic structure. Recently, the present authors found that pure *17b* was very reluctant to polymerize under similar conditions. This is understandable in terms of a smaller enthalpy change from *17b* to its polymer compared with that for *17a*. In the monomeric states, *17b* is less strained than *17a* on account of the equatorial orientation of the bromine atom in the former, whereas in the polymeric states, the polymer from *17b* is energetically less stable than that from *17a*, because the former takes a conformation in which the bromine atom occupies the axial position. Its flipped conformation would be even more unstable, because the stabilization by the anomeric effect is lost, in addition to the axial orientation of the methylene group.

18

3,6,8-Trioxabicyclo[3.2.1]octane *18*, which was prepared from 1,2-chloroethyl-ideneglycerol, was polymerized with a variety of cationic initiators[11]. Phosphorus pentafluoride in dichloromethane gave optimum results. The polymer obtained in quantitative yield showed inherent viscosities of 0.56–0.80 dl/g. The ^{1}H-NMR spectrum of the polymer shows the acetal proton signals at δ 4.76 (equatorial) and 4.67 (axial) ppm. The ratio of equatorial acetal proton to axial is approximately two, indicating that the propagation step in the polymerization of *18* has some S_N1 character. In this case, the ratio of equatorial to axial acetal proton does not change as the inherent viscosity increases from 0.56 to 0.80 dl/g.

2. 6,8-Dioxabicyclo[3.2.1]octenes

19   *20*

Cationic polymerization of 6,8-dioxabicyclo[3.2.1]oct-3-ene *19* and 6,8-dioxabi-
cyclo[3.2.1]oct-2-ene *20* was studied by Okada *et al.*[25, 26]. These monomers contain
a 1,3-dioxolane ring and an olefinic bond which may participate in the cationic polymerization. Therefore, it would be of theoretical interest to clarify the relative reactivities of these two functional groups in the cationic polymerization, in relation to
the copolymerization behavior of cyclic acetals with vinyl monomers. Completely or
nearly completely soluble polymers were obtained from *19* and *20* at −78 °C, even
at high conversions. With the rise in the polymerization temperature, the methylene
chloride-insoluble fraction increased and practically only insoluble polymer was
formed at 0 °C. With higher initial monomer concentration, however, cross-linking
reaction took place appreciably even at −78 °C. The methylene chloride-soluble
polymer of *19* having number average molecular weight of several thousand is a hygroscopic white powder, soluble in a wide variety of solvents. It softens at 72–90 °C,
and begins to decompose in air above 140 °C to form a dark brown insoluble material
with concomitant generation of a small amount of *19*.

21 H_2 Pt   *2*

22

The catalytic hydrogenation of the polymer *21* led to the saturated polymer *22*
which was identical with that produced directly from *2*. This fact, along with [1]H-
and ^{13}C-NMR spectral data, proves that the cationic polymerization of *19* proceeds,
without isomerization of its dihydropyran ring, through the selective cleavage of the
C^5–O^6 bond to yield a polyacetal having a 5,6-dihydro-2H-pyran ring in its repeating
unit[25].

The polymer *21* contains a reactive olefinic linkage in its repeating unit, and can
be modified chemically in various manners. In particular, it is expected that the polymer can be used as a versatile precursor for the chemical synthesis of polysaccharide

analogues. Chemical transformation of *21* to polysaccharide analogues having two hydroxyl groups, one each at 3- and 4-positions (2- and 3-positions in the terminology of carbohydrate chemistry), in its tetrahydropyran unit was achieved in two different ways. One is the epoxidation of the olefinic bond with m-chloroperbenzoic acid or hydrogen peroxide-benzonitrile followed by the alkaline hydrolysis of the resultant epoxy group[27]. The other is the oxidation of the olefinic bond with osmium tetroxide-hydrogen peroxide[28]. The former method leads to the structural unit in which two hydroxyl groups are *trans*-diequatorially or -diaxially oriented, whereas the latter method leads to the structural unit in which two hydroxyl groups are *cis*-oriented.

Poly(*trans*-3,4-dihydroxytetrahydropyran-6,2-diyloxymethylene) *24* showed complex acetal proton NMR signals in the region of δ 4.5–5.2 ppm. This is an indication that there are different kinds of structural units along the main chain. From the estimation of conformational free energies for the possible structural units, along with the reported chemical shifts for the relevant mono- and polysaccharides, it was concluded that there were three structural units with different configurations. Such a complicated structure of the polymer may arise from nonstereospecificity in the epoxidation of the dihydropyran ring and the subsequent alkaline hydrolysis of the epoxide[27].

The newly synthesized polymer *24* having two hydroxyl groups per its repeating unit is expected to display useful properties similar to, or differing from, those of natural carbohydrate polymers. Figure 3 shows the water absorption data for *24*, natural dextran *6*, and poly(tetrahydropyran-2,6-diyloxymethylene) *3*. Interestingly, the water absorption behavior of the dihydroxyl polymer *24* is very similar to that of dextran having three hydroxyl groups per its repeating unit. Water absorption of the former is slightly lower than that of dextran in the range of low and intermediate relative humidities, but conversely, it is higher than that of dextran at high relative humidity. The polymer *3* which has no hydroxyl group scarcely absorbs water except at the highest relative humidity. It is conceivable that the high water absorption observed for the polymer *24* arises mainly from its less stereoregular and therefore disordered structure as revealed by NMR: It brings about incomplete intra- and intermolecular hydrogen bondings, thus making water molecules highly accessible to the polymer chain[27].

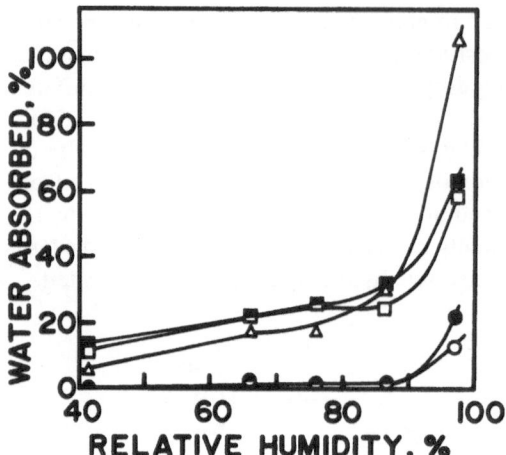

Fig. 3. Water sorption isotherms of poly(trans-3,4-dihydroxytetrahydropyran-6,2-diyloxy-methylene) *24* and its related polymers at 20 °C.
■, dextran *6*, Mn = 76000; □, *6*, Mn = 1100; △, *24*, Mn = 1600; ●, *3*, Mn = 360,000; ○, *3*, Mn = 1700[27)]

3. 2,6-Dioxabicyclo[2.2.2]octane

27   *28*

Hall and his co-workers[15)] prepared 2,6-dioxabicyclo[2.2.2]octane *27* starting from dimethyl malonate and acrolein, and polymerized it with a variety of Lewis and protonic acids. The homopolymers *28* possessed inherent viscosities ranging from 0.13 to 1.1. Clear coherent films of polymer with $\eta_{inh} \geqslant 0.4$ could be cast from chloroform solution. Stereoregular propagation by S_N2 displacement on the bicyclic oxonium ion occurred at low temperatures using fluoro acids as initiators. Stereorandom propagation by S_N1 reaction of an intermediate carbonium ion was observed at 28 °C with methanesulfonic acid or trifluoroacetic acid initiation.

29   *30*

$$\Delta G° = -2.1 \text{ kJmol}^{-1}$$

$$K = \frac{30}{29} = 2.3$$

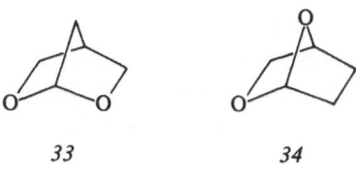

$$\Delta G° = -5.4 \text{ kJmol}^{-1}$$

$$K = \frac{32}{31} = 9.1$$

The difference in conformational energy between the two conformers of the trans-isomer is only 2.1 kJmol^{-1}, resulting in an equilibrium mixture of 30% of *29* and 70% of *30*. Therefore, polymerization of *27* to a *trans*-1,4-tetrahydropyranoside polymer by direct displacement on the trialkyloxonium ion by monomer would be detected by ^{1}H-NMR as an axial acetal proton to equatorial acetal proton ratio $(H_{ax})/(H_{eq})$ of 2.3. The most stereoregular polymer that obtained using 0.2 mol% of SiF$_4$ at -78 °C possessed an $(H_{ax})/(H_{eq})$ ratio of 3.0.

The conformers of the *cis*-isomer have a calculated free energy difference of 5.4 kJmol^{-1} in favor of *32*, which represents an equilibrium mixture of 10% of *31* and 90% of *32*. A carbonium ion as propagating species suffering attack equally from either side to give equal amounts of *cis*- and *trans*-units would afford a calculated $(H_{ax})/(H_{eq})$ ratio equal to 0.67. In fact, several experiments carried out at -20 and $+28$ °C gave, on average, an $(H_{ax})/(H_{eq})$ value of 0.59. Effect of reaction variables on $(H_{ax})/(H_{eq})$ values of the polymer suggests that four factors (polymerization temperature, nature of the catalyst, catalyst concentration, and trace impurities) can influence the stereochemistry of propagation[15].

4. 2,6- and 2,7-Dioxabicyclo[2.2.1]heptanes

33	*34*

Recently, 2,6-dioxabicyclo[2.2.1]heptane *33* and 2,7-dioxabicyclo[2.2.1]heptane *34* were synthesized, and their rates of acidic solvolysis were compared[29a]. Table 1 summarizes the data, along with those for relevant bicyclic acetals determined in a similar manner. Tremendous differences in reactivity among these compounds are noteworthy. The approximate relative reactivities under these conditions spun more than five powers of ten. These differences are attributable to ring strain and anomeric effect.

Polymerization of *33* occurred very readily with a variety of cationic initiators. Protonic acids gave high conversion to soluble polymer which was of moderate molecular weight ($\eta_{inh} = 0.35$), liquid (rubber), and consisted of a 60:40 mixture of iso-

Table 1. Relative reactivities of acidic solvolysis of several bicyclic acetals[a]

Acetal	Relative reactivity	Ref.
2,6-Dioxabicyclo[2.2.1]heptane *33*	6.9×10^5	29)
1,3,3-Trimethyl-2,7-dioxabicyclo[2.2.1]heptane *35*	2.1×10^5	30)
2,7-Dioxabicyclo[2.2.1]heptane *34*	2.5×10^4	29)
2,6-Dioxabicyclo[2.2.2]octane *27*	2.5×10^3	29)
6,8-Dioxabicyclo[3.2.1]oct-3-ene *19*	4.3×10^2	31)
6,8-Dioxabicyclo[3.2.1]octane *2*	7.7	15)
Dimethyl acetal	1	15)

[a]   Conditions: temperature, 35 °C, dichloroacetic acid catalyst; solutions are initially 1.25 M
     in acetal in solvent 0.6 ml acetone-d_6 and 0.2 ml D_2O: rates followed by monitoring acetal
     proton NMR intensity as a function of time.

mers (probably *trans*:*cis*) detected by ^{1}H-NMR. Cationic polymerization of *34* also
occurred readily. Again, high conversions to moderate molecular weight liquid polymer were obtained. At −78 °C, the polymer contained exclusively tetrahydrofuran
links: at a higher temperature, several initiators gave significant amounts of tetrahydropyran links[29a].

The polymerization of 1,3,3-trimethyl-2,7-dioxabicyclo[2.2.1]heptane *35* was
carried out in methylene chloride, toluene, and 1-nitropropane at temperatures between −78 and 0 °C[32]. Boron trifluoride etherate, triethyloxonium tetrafluoroborate, antimony pentachloride, and iodine were used as initiators. Irrespective of
the solvents and initiators employed, the products obtained at 0 °C were white
powders with melting points of 50–55 °C, while those obtained at lower temperatures were sirups. The number average molecular weight of the unfractionated products ranged from 400 to 600. The molecular weight distribution of the oligomers
prepared at 0 °C was broad, in contrast to the relatively narrow distribution of those
obtained at −40 °C.

There are two possible ways for the ring opening of *35* in the polymerization:
The C^1–O^7 bond cleavage leads to the formation of a substituted tetrahydropyran
ring *36* in the polymer chain, while the C^1–O^2 bond cleavage produces a substituted
tetrahydrofuran ring *37*. Product analysis of the acid-catalyzed hydrolysis of the

polymer would not be useful for the elucidation of the polymer structure, because the hydrolysis products (hemiacetals) can be interchangeable through ring-chain equilibrium. Therefore, in order to know which of the C^1-O^7 and C^1-O^2 bonds in 35 is cleaved, preferentially or exclusively, in the polymerization, the hydrogenolysis of 35 with a mixed reagent of lithium aluminum hydride and aluminum chloride was carried out in diethyl ether. The reaction products were found by gas chromatography to consist of two components. The major component which amounted to more than 90% of the total reaction products was identified as 3-hydroxy-2,2,6-trimethyltetra-hydropyran 38 by NMR, IR, and mass spectroscopy, the minor component being 2-(2-hydroxy-2-propyl)-5-methyltetrahydrofuran 39[32].

The preponderance of 38 can be taken as a strong indictaion, but not as a definite proof, for the preferable cleavage of the C^1-O^7 bond over the C^1-O^2 bond in the cationic polymerization of 35. In this connection, it is interesting to note that 1,4-anhydro-2,3,6-tri-O-methyl-D-galactose and 1,4-anhydro-2,3-di-O-methyl-L-arabi-nose possessing the identical skeleton as that of 35 undergo polymerization under the influence of Lewis acid initiators to yield amorphous polymers with a mixture of anomeric linkages and both pyranose and furanose rings[33]. The chain propagation proceeds in each case with nearly equal amounts of five- and six-membered ring-opening leading to a mixture of pyranosidic and furanosidic units in the polymer backbone.

The abnormally low molecular weight of the polymerization products of 35 irrespective of conversions doubtlessly implies the frequent occurrence of monomer transfer during polymerization. End group analysis of the polymer showed that the total number of the hydroxyl groups and carbon-carbon double bonds per oligomer molecule ranged from 1.5 to 1.8 regardless of the polymerization conditions. This finding, along with IR data indicating the presence of trisubstituted olefinic linkage but not of end-methylene groups, suggests that the monomer transfer involves the migration of a proton from the methylene group adjacent to the positively charged carbon atom of the growing ion 40 to the oxygen atom of the monomer[32].

A similar proton transfer from a growing chain end unit to give an olefinic linkage was observed in the cationic polymerization of 2-tert-butyl-7-oxabicyclo[2.2.1]-heptane, although the proton liberated did not initiate the polymerization and hence this process was actually a termination[34].

40   35   41   42

5. 7,9-Dioxabicyclo[4.3.0]nonanes

Attempted polymerization of *cis*-7,9-dioxabicyclo[4.3.0]nonane *43* with phosphorus pentafluoride as initiator at temperatures ranging from −25 to 0 °C provided only a cyclic dimer *44* in high yield[35]. Under similar conditions *trans*-7,9-dioxabicyclo-[4.3.0]nonane *45* polymerized almost instantly to polymer *46* with number average molecular weight of several thousands, along with a small amount of a cyclic dimer *47*.

43   44

45   46   47

The difference in behavior of the *cis*- and *trans*-isomers *43* and *45* is ascribed mainly to the greater ring strain in the *trans*-isomer which results in a more favorable free energy change for its polymerization. This result is similar to that reported for the polymerization of *cis*- and *trans*-8-oxabicyclo[4.3.0]nonane *48* and *49*[36]. The molecular weights of the polymers of the bicyclic ether *49* were, however, much higher than those of the polymers of the bicyclic acetal *45*.

48   49

The bicyclic acetals *43* and *45* can be regarded as 4,5-disubstituted-1,3-dioxolanes. In connection with their polymerizabilities, it is interesting to note here that *cis*-4,5-dimethyl-1,3-dioxolane has a slightly greater tendency to polymerize than its *trans*-counterpart[22, 37]. The polymerization of *45* is an equilibrium reaction and the system is completely reversible. From the temperature dependence of the equilibrium monomer concentrations in toluene-d_8 at temperatures between 25 and 105 °C, the following thermodynamic parameters for the polymerization were obtained.

$\Delta H_{ss}^\circ = -24.1 \pm 0.8$ kJmol^{-1}

$\Delta S_{ss}^\circ = -61.6 \pm 1.8$ Jmol^{-1} K^{-1}

$T_c^\circ = 118$ °C

These values may be compared with the values obtained for the polymerization of 1,3-dioxolane in benzene ($\Delta H_{ss}^\circ = -15.1$ kJmol^{-1} and $\Delta S_{ss}^\circ = -58.5$ Jmol^{-1}K^{-1})[38]. The decrease in enthalpy is greater for the polymerization of *45* and this reflects the greater strain in the dioxolane ring when it is fused in *trans*-position to a six-membered ring. On the other hand, the entropy change is comparable for the two compounds, which is to be expected, because the same ring size is being opened in the polymerization[35].

III. Polymerization of Bicyclic Lactones

The influence of ring size and substitution on the polymerizability of lactones has been reviewed by Hall and Schneider[39] about two decades ago. The general qualitative observations are:

i) Four-, seven-, and eight-membered lactones polymerize, whereas five-membered lactones do not.

ii) The polymerizability of six-membered lactones depend on the substituents.

iii) Alkyl or aryl substituents on a ring always decreases polymerizability.

In contrast to the fact that cyclic acetals can be polymerized only by cationic initiators, lactones undergo polymerization both cationically and anionically, and therefore a wide variety of initiators including coordinated catalysts can be used. In this section, the polymerization of bicyclic lactones is described, although only a limited number of papers on this subject have been published.

1. 2-Oxabicyclo[2.2.2]octan-3-one and 6-Oxabicyclo[3.2.1]octan-7-one

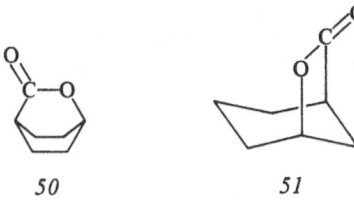

50　　　　　*51*

2-Oxabicyclo[2.2.2]octan-3-one *50*, which is readily prepared by hydrogenation of p-hydroxybenzoic acid followed by dehydration[40], can be polymerized by heating with phosphoric acid or sodium hydride to give a polyester[23]. In contrast, 6-oxabicyclo[3.2.1]octan-7-one *51* polymerizes sluggishly to provide a low molecular weight, solvent sensitive polymer[23]. Such difference in the tendency to polymerize indicates that there is an appreciable strain (H—H repulsion) in the bicyclo[2.2.2]octane system having a cyclohexane ring of a boat form, whereas there is little strain in the bicyclo-[3.2.1]octane system consisting of a chair cyclohexane fused to a cyclopentane ring[23].

Carothers[41] suggested that polymerizability and rates of hydrolysis of cyclic monomers should be parallel. However, this suggestion is not generally valid, because polymerizability of a lactone is closely related to strain in the ring, whereas the ring is not broken in the rate determining step of alkaline hydrolysis. In fact, the rate constant of alkaline hydrolysis of *50* is one order of magnitude lower than that of *51*[42].

2. 2-Oxabicyclo[2.1.1]hexan-3-ones

	R	R'
a	H	H
b	CH_3	H
c	H	CH_3
d	CH_3	CH_3
e	CH_3	CF_3

52(a–e)

Hall and his co-workers[43] synthesized several 2-oxabicyclo[2.1.1]hexan-3-ones from the corresponding 3-chlorocyclobutanecarboxylic acids. These monomers polymerized readily when heated with a variety of basic or acidic initiators. Some of the results of the polymerization are listed in Table 2.

High molecular weight, rather high melting polymers were obtained and could be melt pressed or cast into films or spun into fibers. These polymers were soluble

Table 2. Polymerization of 2-oxabicyclo[2.1.1]hexan-3-ones[43]

Lactone, g		Xylene, ml	Initiator	Temp., °C	Time, hr	Yield, %	η_{inh}[a]	M.p.,[b] °C
52a	1.0		Ti(OiPr)$_4$, 0.01 ml	150	1.8	98	0.26	236
52a	1.0		NaOMe, 0.01 g	150	1.8	80	0.41	242
52b	1.0		NaOMe, 0.01 g	150	16	53		
52b	1.0		Et$_3$Al, 0.01 ml	150	2	26	0.23	245
52c	2.0	1.7	NaK	45	1.7	85	3.33	
52c	2.0	3.0	Ph$_3$CNa	50	20	29	2.55	
52d	2.0	2.0	NaK	42	20	100	3.43	
52e	0.9		BF$_3$OEt$_2$	68	1.5	55	0.52	

[a] CF_3CO_2H, 0.1%, 30 °C.
[b] DTA.

in acidic solvents such as trifluoroacetic acid and in halogenated solvents such as chloroform. They conclude that the 2-oxabicyclo[2.1.1]hexan-3-ones are a strained group of lactones resembling β-lactones rather than γ-lactones in their ability to polymerize. Up to two substituents on the ring do not decrease the tendency to polymerize, again like the β-lactones. Copolymerization of *52d* and pivalolactone with sodium methoxide as initiator gave a copolymer having an inherent viscosity of 0.19 dl/g and a melting point of 190–192 °C. This copolymer could be melt pressed at 170 °C into a clear film.

3. 6,8-Dioxabicyclo[3.2.1]octan-7-one

Tamura *et al.*[24] described only briefly in their paper on dihydropyran derivatives that 6,8-dioxabicyclo[3.2.1]octan-7-one *53* and its methyl derivative underwent polymerization in the presence of a large amount of boron trifluoride etherate to give polymers with molecular weights of several hundreds.

As a continuation of the series of studies on the ring opening polymerization of bicyclic compounds having a tetrahydropyran ring, Okada *et al.*[44, 45] investigated the polymerization of *53* in more detail. Interestingly, it was found that the cationic polymerization of *53* gave rise to cyclic oligomers *54* with degrees of polymerization of two, four, and six, or polymers with molecular weights up to 2000, depending on the reaction conditions.

n = 2, 4, and 6

53 *54*

The effect of the temperature on the polymerization of *53* in methylene chloride is presented in Table 3. The upper half of the data in the table shows the temperature effect on the products in the initial stage of the reaction, and the lower half is that for the middle to final stages of the reaction. Obviously there is a drastic change in the reaction products between −20 and −30 °C: Below −30 °C, the cyclic dimer is the predominant or even sole product after the reaction of 48 hours, while above −20 °C, the low molecular weight polymer is exclusively formed. The cyclic oligomers once formed in the initial stage of the reaction are converted to the polymer in the later stage of the reaction above −20 °C.

Time dependence of the reaction products can be seen more clearly in the time-yield curves of oligomerization in methylene chloride at −40° (Fig. 4). The yield of mixture of the cyclic tetramer and hexamer (mostly the latter), passed through a maximum value of about 40% and then decreased to nearly zero after 48 hours. On the other hand, the yield of the cyclic dimer increased rather sigmoidly with reaction time.

Table 3. Effect of temperature on the polymerization of 6,8-dioxabicyclo[3.2.1]octan-7-one in methylene chloride[a] [45)]

		Yield[b], %			
Temp., °C	Time, hr	Cyclic dimer	Cyclic tetramer + cyclic hexamer	Polymer	$\overline{M}n^{c}$
0	0.3	0	0	1	
−10	0.3	8	0	1	
−20	0.3	9	12	0	
−30	0.3	20	11	0	
−40[d]	0.3	3	10	0	
0	48	0	0	79	1900
−10	48	0	0	76	1300
−20	48	0	0	56	1300
−30	48	62	~0	4	
−40[d]	48	77	~0	0	

a   Monomer, 4.0 g; BF_3Et_2O, 0.5 mol% to monomer; solvent, CH_2Cl_2, 4.0 ml.
b   Determined by NMR or liquid chromatography.
c   Determined by vapor pressure osmometry in benzene at 37 °C.
d   Initiator, 1.0 mol% to monomer.

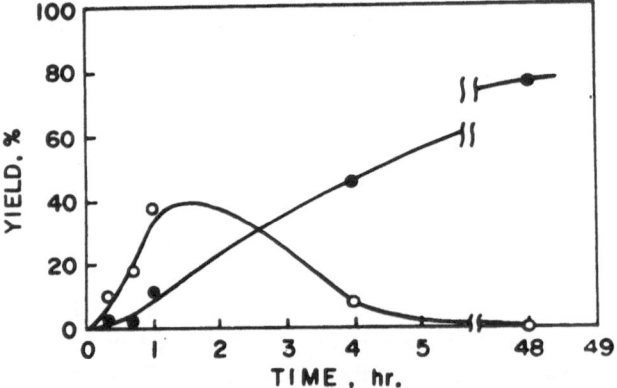

Fig. 4. Time-yield curves of the oligomerization of 6,8-dioxabicyclo[3.2.1]octan-7-one. Monomer, 4 g; CH_2Cl_2, 4 ml; BF_3Et_2O, 1 mol% to monomer; temp., −40 °C; •, cyclic dimer; ○, cyclic tetramer plus cyclic hexamer[45)]

Such products were observed more or less in the polymerization with other initiators ($SnCl_4$, $SbCl_5$, CH_3COBF_4, Et_3OBF_4, and $EtOSO_2F$). These phenomena are inconsistent with a usual equilibrium process if, as expected, the enthalpy of the polymerization is negative. Therefore, solubility and kinetic factors must play an important role.

Table 4 summarizes the most suitable reaction conditions for the selective preparation of cyclic dimer, tetramer, and hexamer from *53*. It must be pointed out

Table 4. Selective oligomerization of 6,8-dioxabicyclo[3.2.1]octan-7-one at −40 °C

Monomer, g	Solvent[a], ml		Time, hr	Conversion, %		
				Dimer	Tetramer	Hexamer
4[b]	MC	4	48	77	0	0
2[c]	CF	2	24	0	91	0
2[b]	NP	4	24	2	1	56

[a] MC, methylene chloride; CF, chloroform; NP, 1-nitropropane.
[b] BF_3Et_2O, 1 mol% to monomer.
[c] BF_3Et_2O, 10 mol% to monomer.

that the cyclic hexamer was predominantly produced in 1-nitropropane under the specified conditions given in Table 4, but it was converted gradually into the cyclic dimer when the reaction mixture was allowed to stand for a very long time. All the reaction systems listed in Table 4 become heterogeneous as the reaction proceeds. Comparison of the data in Table 4 with the solubility data of the cyclic oligomers in three different solvents (Table 5) reveals that solubility is a main factor controlling the selective formation of the cyclic oligomer of a particular ring size[46].

Table 5. Solubilities of cyclic oligomers of 6,8-dioxabicyclo[3.2.1]-octan-7-one at − 40 °C[46]

Solvent	Solubility (g/100 ml)		
	Dimer	Tetramer	Hexamer
Methylene chloride	0.33	19	36
Chloroform	1.9	0.88	40
1-Nitropropane	0.17	5.1	1.7

The cyclic dimer, tetramer, and hexamer can be crystallized in acetonitrile, and also in chloroform (the former two oligomers). X-ray analysis of the crystals of the cyclic dimer[47] disclosed that it consisted of a pair of different enantiomers of 53 and that all of the four substituents attached to the two tetrahydropyran rings occupied the axial position as illustrated in Fig. 5[45]. Such a conformation is in

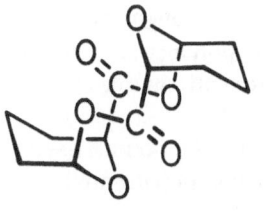

Fig. 5. Structure of the cyclic dimer of 6,8-dioxabicyclo-[3.2.1]octan-7-one[45]

accord with the ^{1}H- and ^{13}C-NMR data of the cyclic dimer in deuterochloroform
solution. X-ray analysis of the crystals of the cyclic tetramer and hexamer has not
yet been accomplished. ^{13}C-NMR spectra of these oligomers consist of only five
peaks, one of which is the overlapped signal of two different methylene carbons in a
tetrahydropyran ring. Such simple spectra suggest that these oligomers are not a
mixture of two or more geometrical isomers, but a single compound with a high
molecular symmetry. The ^{1}H- and ^{13}C-NMR spectra of the cyclic tetramer and
hexamer can be explained most satisfactorily in terms of the conformation in which
the acetal hydrogens, like those in the cyclic dimer, occupy the equatorial position
of the tetrahydropyran ring whereas the methine hydrogens adjacent to the carbonyl
group are located in the axial position, unlike those in the cyclic dimer.

It is not an unusual but a rather common phenomenon that cyclic oligomers of
various ring sizes are formed in the cationic polymerization of a variety of cyclic
monomers[48]. The formation of cyclic oligomers has been interpreted generally in
terms of back-biting reaction of the growing chain ends. In the present case, only
three kinds of cyclic oligomers (DP = 2, 4, and 6) were formed. Hence it is very
peculiar, and difficult to explain, that competitive formation of neither polymer nor
cyclic oligomers of other sizes were observed, if the cyclic oligomers were formed by
a simple back-biting mechanism. Probably, it would appear that the growing chain
takes a particular conformation suitable for the formation of the even-numbered
cyclic oligomers under the reaction conditions used. The alternating arrangement of
a 2,6-disubstituted tetrahydropyran ring and an ester linkage of the growing chain
seems to be responsible for the ease of its taking the conformation favorable for the
ring closure, because as described in the foregoing reaction, *2*, having the same skeleton as that of *53*, polymerizes readily to a high polymer under similar conditions[12, 13],
whereas 5-methyl-1,3-dioxolan-4-one which has a partial structure of *53* forms
neither cyclic oligomer nor linear polymer[49].

An alternative explanation for the exclusive formation of the even-numbered
cyclic oligomers would be that they are produced by the polymerization of the cyclic
dimer. However, the cyclic dimer isolated from the reaction mixture was found not
to polymerize under the reaction conditions used for the oligomerization. Therefore,
if this concept were correct, another cyclic dimer with a different configuration and
higher reactivity must have been formed during the reaction. Its presence, however,
has not yet been verified by experiment.

The structure of the low molecular weight polymer obtained at higher temperature must be mentioned briefly. It is intriguing that the structure of the polymer is
not polyester as expected from the analogy of the polymerization of 2^{10-13}. The IR
spectrum of the polymer shows two strong carbonyl absorptions at 1790 and
1740 cm^{-1} which are characteristic of an acid anhydride structure. Furthermore,
the acetal methine proton of the polymer exhibits its NMR signal at δ 5.6 ppm which
is significantly different from the corresponding signals of the cyclic oligomers
(δ 6.36–6.58 ppm). These spectroscopical data suggest that the polymer contains an
anhydride-acetal structure *55*[44]. Such a structure may arise from alternating sequences
of alkyl-oxygen and acyl-oxygen fissions of the lactone ring in *53*.

Since the pioneering work by Pedersen[50], heteromacrocyclic compounds have
attracted much attention in various fields because of their high complexing abilities.

55

Numerous heteromacrocycles containing oxygen, nitrogen, sulfur, phosphorus and so on have been synthesized, and their complexing behaviors have been investigated in detail. The thirty-membered cyclic hexamer of *53* bears a structural resemblance to a naturally occurring antibiotic nonactin *56* which is a thirty two-membered cyclic oligoester having a tetrahydropyran ring and an ester linkage in its repeating unit and acts as a molecular carrier for metal cations, especially for potassium ion[51].

56

It is expected, therefore, that the cyclic hexamer also exhibits a characteristic tendency to complex with cations. In fact, the addition of an acetonitrile solution of metal thiocyanates to a solution of the cyclic hexamer in the same solvent shifted the carbonyl absorption to a lower wave number[46, 52]. The shift values depended upon the kind of metal ions present, and the largest shift value of 40 cm^{-1} was observed for barium thiocyanate (molar ratio of Ba^{2+}/hexamer = 10). In addition to the shift of the carbonyl absorption, the intensities of the $C-O-C$ stretching vibrations around 1200 cm^{-1} varied appreciably.

Metal thiocyanates generally do not dissolve in chloroform, but they dissolve in it, although gradually, in the presence of the cyclic hexamer. Alternatively, an acetonitrile solution of the cyclic hexamer and metal thiocyanates is evaporated to dryness under reduced pressure, and subsequently chloroform is added to the residue to produce a clear solution, unless the molar ratio of metal ion to the cyclic hexamer greater than 2. Table 6 summarizes ^{1}H-NMR shift changes of the cyclic hexamer when it complexed with an equimolar amount of metal and ammonium thiocyanates in deuterochloroform. The acetal methine proton (a) shows, in most cases, a small upfield shift, whereas the methine proton (b) adjacent to the carbonyl group exhibits relatively large downfield shifts. The most pronounced shift was observed for the barium salt, corresponding to the largest shift of the infrared carbonyl absorption.

Proton noise decoupled ^{13}C-NMR spectra of equimolar mixtures of the cyclic hexamer and metal thiocyanates showed that the signals of the carbonyl carbon and two methine carbons gave downfield shifts upon the addition of metal thiocyanates, while those of the three methylene carbons of the tetrahydropyran ring gave upfield

Table 6. ^{1}H-NMR chemical shift changes of the cyclic hexamer of 6,8-dioxabicyclo[3.2.1]-octan-7-one upon interaction with thiocyanate salts in deuteriochloroform[a]

Thiocyanate	$\Delta\delta$, ppm		
	a	b	c
LiSCN · H$_2$O	0.01	0.08	0.02
NaSCN	−0.02	0.16	0.02
KSCN	−0.05	0.10	0.00
Ca(SCN)$_2$ · 3 H$_2$O	−0.05	0.19	0.00
Ba(SCN)$_2$ · 2 H$_2$O	−0.05	0.43	0.01
NH$_4$SCN	−0.03	0.09	0.01

[a] Cyclic hexamer : thiocyanate salt = 1 : 1 (mol); concn., 3%; 100 MHz; room temp.; positive and negative values refer to downfield and upfield shifts, respectively.

shifts. The large shift values for the carbonyl carbon ranging from 1.02 to 3.05 ppm demonstrate that there is an interaction primarily between the carbonyl oxygens and the cations of the salts. The unusually large shift values observed in the case of barium thiocyanate strongly suggest that there may be a specific interaction between the hexamer and the barium cation. Solvent effect on the chemical shifts is not uniform. Presumably, this is an indication that the cyclic hexamer, when it interacts with thiocyanate salts, takes somewhat different conformations depending on the kinds of cations of the metal thiocyanates and solvents used.

4. 3,8-Dioxabicyclo[3.2.1]octan-2-one

57 58

As a first step in the preparation of polymers patterned after the repeating unit of nonactin 56, Moore and Kelley[53] synthesized 3,8-dioxabicyclo[3.2.1]octan-2-one 57 and its corresponding polyester 58. The monomer was prepared from 5-hydroxy-methylpyran-2-carboxylic acid in overall yield of 20%. It was heated with a catalytic amount of tert-butoxytitanate under nitrogen for 3 hr at 100 °C. The temperature was then raised to 170 °C over 2 hr and maintained at 170 °C for 1 hr. A sticky solid having an inherent viscosity of 0.18 was obtained.

Although it is unlikely that the tetrahydrofuran ring would open under these conditions of polymerization, the polymer was hydrolyzed in 0.2 M sodium hydroxide solution, in order to confirm its structure. Clear colorless liquid was obtained after acidification followed by esterification with diazomethane. Its IR and NMR data compares exactly to that obtained from *59* which was prepared from the neutral hydrolysis of *57* and esterification of the resultant acid with diazomethane. Since the apparent sole product obtained from hydrolysis of the polymer was *59*, they conclude that *58* represents the correct structure for this polymer.

59

IV. Polymerization of Bicyclic Oxalactam

1. Bicyclic Lactams

Syntheses of a number of bicyclic lactams and their qualitative polymerizabilities related to the steric strains were described and comprehensively summarized by Hall[23, 54], as simply shown in Table 7.

Table 7. Polymerizabilities of bicyclic lactams[54]

Lactam structural formula	Name	Polymerizability
	2-Azabicyclo[2:2:1]heptan-3-one	Predict +
	2-Azabicyclo[2:2:2]octan-3-one	+
	2-Azabicyclo[3:2:1]octan-3-one	–
	6-Azabicyclo[3:2:1]octan-7-one	+
	3-Azatricyclo[3:2:1:0^{4,6}]octan-2-one	–
	2-Azabicyclo[3:2:2]nonan-3-one	+
	2-Azabicyclo[3:3:1]nonan-3-one	Predict –
	2-Azabicyclo[4:2:1]nonan-3-one	+
	6-Azatricyclo[3:2:1:1^{3,8}]nonan-7-one	Predict +
	1-Azabicyclo[3:3:1]nonan-2-one	+

Melting points of the polymers from bicyclic lactams are expected to be appreciably high because of some conformational restriction of the main chain due to the ring structural unit, as seen from the examples[54] presented on page 25.

It has also been known that the polymerization of some bicyclic lactam ethers, for example, 3-aza-10-oxabicyclo[4.3.1]decan-4-one 60 was accompanied by the cleavage of the ether bond[55, 56].

60

2. 8-Oxa-6-azabicyclo[3.2.1]octan-7-one

A new class of bicyclic oxalactam, 8-oxa-6-azabicyclo[3.2.1]octan-7-one 61 was synthesized by intramolecular cyclization of 3,4-dihydro-2H-pyran- 2-carboxamide 62 using p-toluenesulfonic acid as a catalyst in an equivalent mixture of DMF and benzene at 100 °C for 4 hours[57, 58]. Corresponding conversion of 3-cyclohexane carboxamide 63 to bicyclic lactam 64 has never been accomplished[54].

62   61

63   64

61 was obtained as a white waxy solid, bp 114 °C (4 mmHg). Recrystallization from n-hexane yields colorless fine needle-shaped crystals, mp 91–92 °C.

Some of the results of bulk polymerization of 61 by using different anionic catalysts are summarized in Table 8[58]. It was easily polymerized in the presence of alkali metal compounds above 60 °C. The polymerization at 150 °C was too fast to be controlled. The yield and the viscosity number, η_{sp}/c, of the resulting polyamide increased with the reaction time. The initial rate of the polymerization became higher with the size of the countercation, in analogy to the case of anionic polymerization of ε-caprolactam[59]. The rate increased also with raising temperature as shown in Fig. 6[58].

Table 8. Bulk polymerization of 8-oxa-6-azabicyclo[3.2.1]octan-7-one *61*[58)]

61, g	Catalyst	Catalyst, mol%/ monomer	N-Acetylated *61*, mol%/ monomer	Temp., °C	Time, min	Yield, %	η_{sp}/c[a]
0.64	K Pyrdn	1	0	100	10	40	0.61
0.64	K Pyrdn	1	0	100	180	68	1.1
0.64	K Pyrdn	1	0	100	360	82	b
0.64	K Pyrdn	1	0	70	10	16	0.50
0.64	K Pyrdn	1	0	70	60	44	0.77
4.65	K Pyrdn	1	0.2	70	10	72	0.51
0.64	K Pyrdn	1	0	60	60	10	0.40
0.64	K Pyrdn	1	0	50	60	~0	
0.64	K Pyrdn	1	0.2	50	60	1.4	
2.80	NaH	1	0.2	70	10	62	0.43
1.27	n-BuLi	1	0.2	70	60	32	0.26
0.63	NEt₃	10	1	100	720	0	

a   In m-cresol, c = 0.2 g/100 ml, 25 °C.
b   Insoluble in m-cresol.

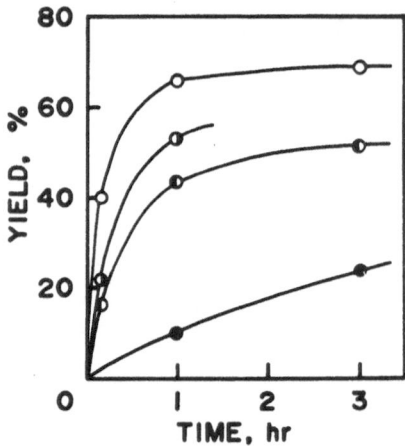

Fig. 6. Time-conversion curves for the bulk polymerization of 8-oxa-6-azabicyclo[3.2.1]octan-7-one. ○, 100 °C; ◑, 80 °C; ◉, 70 °C; ●, 60 °C[58)]

It is well known that the anionic polymerization of lactams in general is markedly facilitated by the addition of N-acetylated lactams[60−62)]. The polymerization of *61* is also accelerated by the addition of its N-acetylated compound *65* (Table 8).

The anionic polymerization of *61* catalysed by potassium pyrrolidonate is characterized by a rapid conversion in THF and DMSO, even at or below room temperatures, as indicated in Table 9[58)].

In THF the reaction system became turbid and gelled as the polymerization proceeded. In DMSO it proceeded in a homogeneous phase. It is also worth noting that *61* is able to polymerize in these solvents even in the absence of activator *65*. The prolonged polymerization resulted in the formation of a sufficiently high molecu-

Table 9. Solution polymerization of 8-oxa-6-azabicyclo[3.2.1]octan-7-one *61*[a] [58]

61, g	Solvent	S/M[b]	N-Acetylated *61*, mol%/ monomer	Temp., °C	Time, h	Yield, %	η_{sp}/c[a]
1.31	THF	4.9	0	50	1	20	0.80
1.30	THF	4.9	0.2	25	1	67	0.53
1.36	THF	4.9	0.2	0	1	15	0.26
1.00	DMSO	2.8	0	50	1	61	0.99
1.39	DMSO	2.8	0.2	50	1	79	0.59
2.65	DMSO	5.6	0	25	2	28	1.70
2.24	DMSO	5.6	0.2	25	1	92	0.64
5.92	DMSO	5.6	0	19	72	56	1.97

[a]   K Pyrdn, 1 mol%/monomer.
[b]   Molar ratio of solvent to monomer.
[c]   In m-cresol, 0.2 g/100 ml, 25 °C.

lar weight polymer. These facts indicate that the anion *66* produced by proton exchange between *61* and pyrrolidonate anion from the catalyst can also easily attack the amide group of another *61* monomer as well as the imide group of *65* as shown in Eqs. (1) and (2).

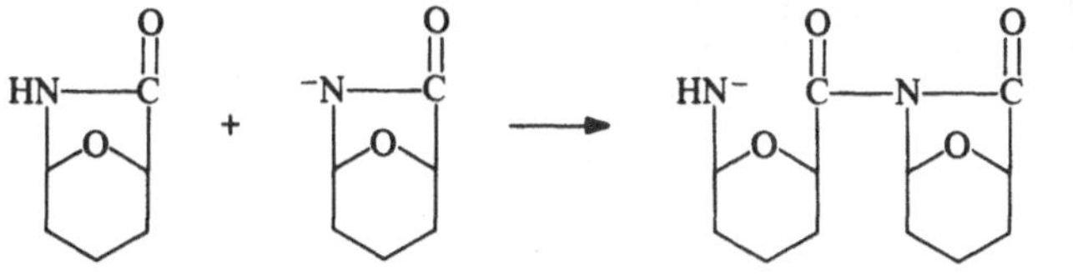

(1)

61   *66*

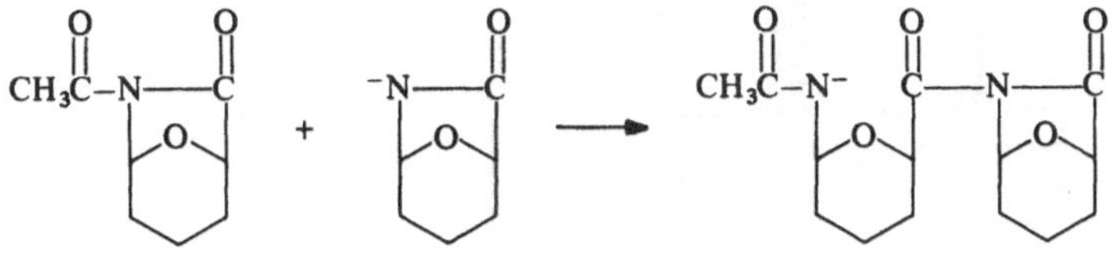

(2)

65   *66*

From the kinetic viewpoint the polymerizability of *61* is considered to be higher than that of ε-caprolactam, which is polymerized usually at temperatures above 135 °C[63, 64]. Thermodynamically, the polymerization of *61* appears to be more favored than that of α-pyrrolidone, for which no polymerization is observed in THF[63-65]. The higher polymerizability of *61* may be attributed not only to its highly strained bicyclic structure but also to the activation of the anion *66* by the separation of the countercation from the amidate anion, under the influence of possible interaction of the cation with the ring ether oxygen. Thus room-temperature polymerization and simultaneous film casting on a glass or a metal plate become feasible ("casting polymerization")[58].

3. Poly(tetrahydropyran-2,6-diyliminocarbonyl)

IR and NMR Analyses. The infrared spectrum of the polymer from *61* shows absorptions in chloroform at 3400 and 3500 (ν(NH)). 3000, 2950, 2920, 2860, 1678 (ν(C=O)), 1528 (δ(NH)), 1440, 1300, 1193, 1145, 1092, 1068, 1030, and 919 cm^{-1} and indicates that the cyclic amide linkage in the monomer was transformed to the linear amide group.

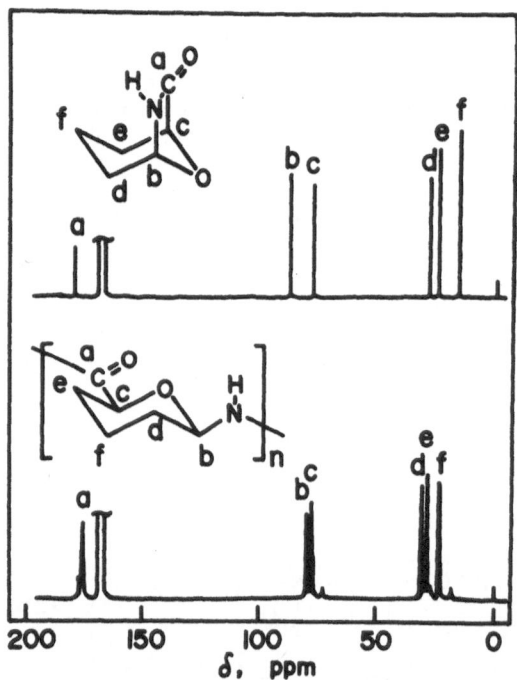

Fig. 7. ^{13}C-NMR spectra of 8-oxa-6-azabicyclo[3.2.1]octan-7-one and its polymer (7% in formic acid). Polymerization: at 70 °C for 10 min in 72% yield; η_{sp}/c 0.51 (in m-cresol, 0.2 g/100 ml, 25 °C)[58]

61   *67*

The ^{13}C-NMR (Fig. 7) and ^{1}H-NMR spectra also support that the polymer is a
novel polyamide, poly(tetrahydropyran-2,6-diyliminocarbonyl) 67. The chemical
shift of peak f in the ^{13}C-NMR spectrum of the polymer is lower by 7 ppm than that
of the monomer which shifts to the higher mangetic field because of the steric effect
of the axial substituents on the tetrahydropyran ring. In addition, the chemical shift
of peak b in the spectrum of the polymer is higher by 8 ppm than that of the monomer which shifts to the lower magnetic field due to the magnetic anisotropic effect
of the ring carbonyl group. These shifts indicate that the amide group is located in
the axial position to the tetrahydropyran ring in the monomer and in the equatorial
one in the polymer. In addition to the six main peaks, there appear some small peaks
in the spectrum, which may result from stereoisomerism of the repeating unit or
branched structures.

Thermal Analysis. The DSC thermogram of the polymer prepared in DMSO at
19 °C has two endothermic peaks at 260—285 and 315 °C due to the fusion and
decomposition, respectively, as shown in Fig. 8.

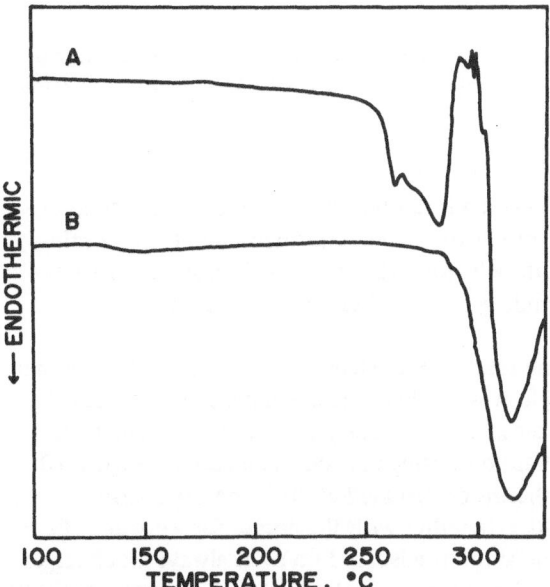

Fig. 8. DSC thermograms of poly(tetrahydropyran-2,6-diyliminocarbonyl) prepared in DMSO
at 19 °C for 72 hr in 56% yield. Heating rate, 10 °C/min. A, original scanning; B, rescanning
after scanning up to 290 °C[58)]

Another transition appeared at about 130 °C on the rescanning after rapid cooling of the molten sample. It seems to be a glass transition point. The polyamide prepared in bulk at 70 °C also melted sharply at 250—260 °C. On the other hand, the
crosslinked polymer, which was prepared in bulk at 100 °C in a high conversion, has
nothing but a broad endothermic curve up to near 300 °C as shown in Fig. 9.

Solubility. Most samples of the resulting polymer were soluble in formic acid,
m-cresol, DMSO, and even chloroform, while the polymer prepared in bulk at elevated

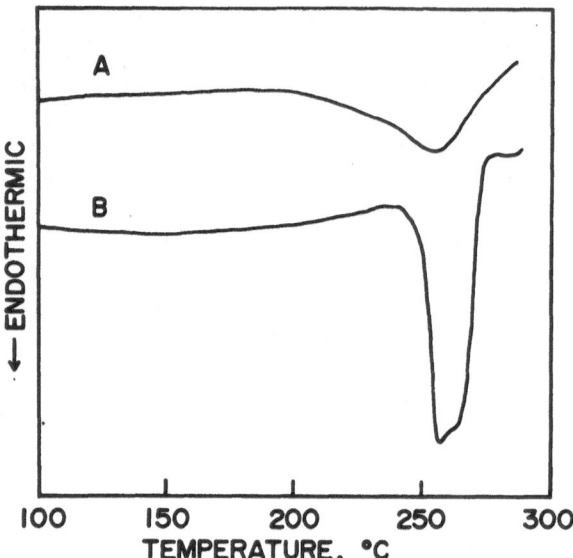

Fig. 9. DSC thermograms of polyamide prepared by bulk polymerization of 8-oxa-6-azabicyclo-[3.2.1]octan-7-one. Heating rate, 10 °C/min. Polymerization: A, at 100 °C for 6 hr in 82% yield; B, at 70 °C for 3 hr in 52% yield[58]

temperatures at a high conversion was often insoluble even in m-cresol. These solubility behaviors reflect a linear structure of the latter. The uncommonly good solubility of this polyamide in such aprotic solvents as DMSO and chloroform suggests that the hydrogen bonds between amide groups are loosened by the interference from bulky tetrahydropyran rings.

Moisture Sorption Isotherms. The results of an elemental analysis of the polymer, dried at 100 °C until a constant weight was reached, showed a discrepancy from the calculated values, suggesting that water persisted in the polymer. The polymer film is flexible in water but more or less hard on drying at room temperature. Figure 10 represents the moisture sorption isotherms determined at 20 °C on the coarsely ground samples of different polyamides, together with the curves for wool and silk fibers cited from the literature[66]. The amount adsorbed on 67 is always much larger than that of either poly(α-pyrrolidone) or poly(ε-caprolactam) at any relative humidity and is, roughly speaking, comparable to wool and silk. The observed moisture sorption of 42% at 99% relative humidity corresponds to the sorption of more than three molecules of water per a monomer unit in the polymer chain. Such a distinguished hygroscopic property of this polymer may be accounted for by the occurrence, even if partly, of hydrophilic polar microdomains surrounded by hydrophobic nonpolar microdomains, which result from the alternating arrangement of an amide linkage and a tetrahydropyran ring along the chains. Possible molecular arrangements, providing all hydrogen bonds are ideally regularly formed between polymer chains, are illustrated in Fig. 11. In the presence of water hydrogen bond interactions must exist between the water molecules and the amide groups. The ring ether oxygens may also interact with the water molecules.

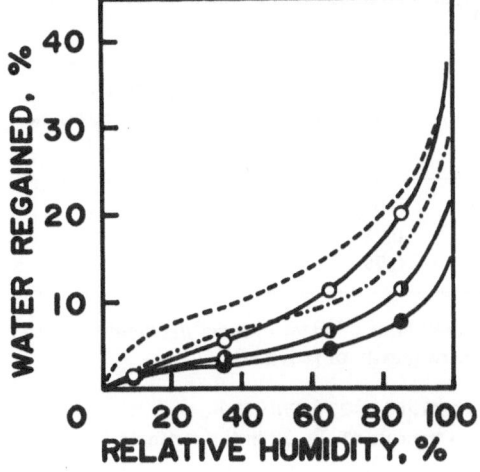

Fig. 10. Sorption isotherms of different polyamides. At 20 °C. ○, *67*; ◑, nylon-4; ●, nylon-6; – – –, wool; –·–, silk[58]

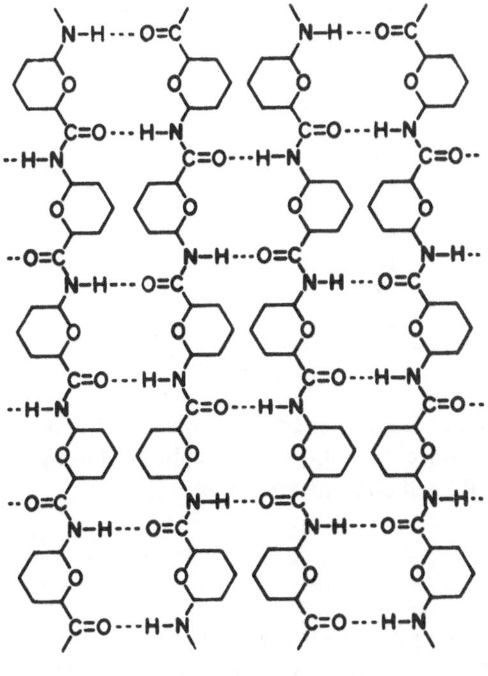

Fig. 11. Possible planar arrangements of poly(tetrahydropyran-2,6-diylimino-carbonyl) molecules provided all hydrogen bonds are ideally regularly formed between the chains[58]

Water Permeation and Solute Separation through the Membrane. The measurements of water permeability of the *67* membranes prepared under different conditions were carried out by using an Amicon Diaflo Cell (effective membrane area, 13.9 cm²) under a pressure of 3 kg/cm² at 25 °C. Some results are listed in Table 10[67]. It is apparent that much higher water absorption and permeability than the cellulosic membrane are characteristic of the *67* membranes prepared by both the casting polymerization and conventional casting.

Table 10. Water permeation through poly(tetrahydropyran-2,6-diyliminocarbonyl) membrane under the pressure of 3 kg/cm^2 at 25 °C[67, 68]

Membrane No.	Thickness, mm	Degree of hydration[a]	Water permeation, $1/m^2hr$
Cellulose[b]	0.05	0.06	1.7
29[c]	0.50	0.78	7.3
C-30[d]	0.38	0.79	8.9

[a] Volume fraction of water in a wet membrane. (Volume of water included)/[(volume of water included) + (weight of dried membrane)/(specific gravity of polymer)].
[b] Visking dialyzer tubing.
[c] Obtained by the casting polymerization in DMSO at 19–22 °C for 72 hr.
[d] Taken out at almost the same concentration as the case of No. 29 in the casting process from 5% solution of polymer in DMSO.

Table 11. Solute rejection in aqueous solution (3 kg/cm^2, 25 °C)[68]

Solute	Mol. wt.	R_0, %
Sodium chloride	58.5	22
Creatinine	113	3
Uric acid	168	14
D-(+)-Glucose	180	8
Vitamine B_{12}	1,355	78
Myoglobin	17,800	88
Alubumin	67,000	94
γ-Globulin	175,000	98

Table 11 presents the rejection of solutes from their aqueous solutions by the 67 membrane with the apparatus mentioned above. Rejection R_0 was obtained using Eqs. (3) and (4) from apparent rejection R_a and concentration factor X.

$$R_a = 1 - (c_2/c_0) \tag{3}$$

$$R_0 = \log[R_a(X - 1) + 1]/\log X \tag{4}$$

where c_0 and c_2 are concentrations of sample solutions before and after permeation, respectively, which are determined by UV spectroscopy or the conductometry[68].

The permeability tests for alkali metal ions in the aqueous solution were also conducted. When an aqueous salt solution moves to cell 2 through the membrane from cell 1, the apparent diffusion coefficient of the salt D can be deduced from a relationship among the cell volumes V_1 and V_2, the solution concentration c_1 and c_2, the thickness of membrane, and time t[69]. In Table 12, permeabilities of potassium chloride and sodium chloride through the 67 membrane prepared by the casting polymerization technique from the monomer solution in THF or DMSO are compared with each other and with that the permeability through Visking dialyzer tubing. The

Table 12. Permeability of potassium chloride and sodium chloride through poly(tetrahydro-pyran-2,6-diyliminocarbonyl) membrane in aqueous solution at 25 °C[70]

| Membrane No. | Casting polymerization | | | Thickness, mm | $\dfrac{D_{K^+}{}^a}{D_{K^+,V}}$ | $\dfrac{D_{Na^+}{}^a}{D_{Na^+,V}}$ | $\dfrac{D_{K^+}}{D_{Na^+}}$ |
	Solvent	Temp., °C	Time, hr				
Cellulose[b]	–	–	–	0.05	1.0	1.0	1.20
I-95	THF	50	3	0.37	7.0	6.3	1.33
I-100	DMSO	50	4	0.44	7.8	8.1	1.15
9	DMSO	19	72	0.93	9.4	9.2	1.23

[a] Ratio of apparent diffusion coefficient of 67 membrane to the cellulosic membrane.
[b] Visking dialyzer tubing.

apparent diffusion coefficients of both alkali metal ions are much higher in the 67 membrane than in the cellulosic membrane. It is also interesting to note that the diffusion coefficient of potassium ion is more or less higher than that of sodium ion[70].

V. References

[1] Korshak, V. V., Golova, O. P., Sergeev, V. A., Merlis, N. A., Schneer, R. Y.: Vysokomol. Soedin. 3, 477 (1961)
[2] Tu, C. C., Schuerch, C.: J. Polym. Sci. B1, 163 (1963)
[3] Schuerch, C.: Adv. Polym. Sci. 10, 174 (1972)
[4] Schuerch, C.: Acc. Chem. Res. 6, 184 (1973)
[5] Okada, M., Yamashita, Y., Ishii, Y.: Makromol. Chem. 80, 196 (1964)
[6] Firat, Y., Plesch, P. H.: J. Polym. Sci., Polym. Lett. Ed. 13, 135 (1975)
[7] Okada, M., Hisada, T., Sumitomo, H.: Makromol. Chem. 179, 959 (1978)
[8] Dainton, F. S., Ivin, K. J.: Quart. Rev. (Chem. Soc. London) 12, 61 (1968)
[9] Sweet, F., Brown, R. K.: Can. J. Chem. 46, 2289 (1968)
[10] Kops, J.: J. Polym. Sci. Part A-1, 10, 1275 (1972)
[11] Hall, Jr., H. K., Steuck, M. J.: J. Polym. Sci., Polym. Chem. Ed. 11, 1035 (1973)
[12] Sumitomo, H., Okada, M., Hibino, Y.: J. Polym. Sci., Polym. Lett. Ed. 10, 871 (1972)
[13] Okada, M., Sumitomo, H., Hibino, Y.: Polym. J. 6, 256 (1974)
[14] Jackman, L. M., Sternhell, S.: Application of nuclear magnetic resonance spectroscopy in organic chemistry. 2nd edit. New York: Pergamon Press 1969, p. 238
[15] Hall, Jr., H. K., Carr, L. J., Kellman, R., DeBlauwe, Fr.: J. Am. Chem. Soc. 96, 7265 (1974)
[16] Angyal, S. J.: The carbohydrates 2nd Edit. Pigman, W., Horton, D. (eds.). New York: Academic Press 1972, Vol. IA, p. 202
[17] Zachoval, J., Schuerch, C.: J. Am. Chem. Soc. 91, 1165 (1969)
[18] Yokoyama, Y., Okada, M., Sumitomo, H.: Makromol. Chem. 176, 795 (1975)
[19] Yokoyama, Y., Okada, M., Sumitomo, H.: Makromol. Chem. 178, 529 (1977)
[20] Okada, M., Sumitomo, H., Hibino, Y.: Polym. J. 7, 511 (1975)
[21] Plesch, P. H., Westermann, P. H.: J. Polym. Sci. Part C 16, 3837 (1968)
[22] Okada, M., Mita, K., Sumitomo, H.: Makromol. Chem. 177, 2055 (1976)
[23] Hall, Jr., H. K.: J. Am. Chem. Soc. 80, 6412 (1958)
[24] Tamura, A., Nishi, T., Uehara, K., Murata, J.: Kogyo Kagaku Zasshi (J. Chem. Soc. Jpn., Ind. Chem. Sect.) 68, 2271 (1965)

25) Okada, M., Sumitomo, H., Komada, H.: Makromol. Chem. *178*, 343 (1977)
26) Okada, M., Sumitomo, H., Komada, H., Yamazaki, Y.: to be published
27) Okada, M., Sumitomo, H., Komada, H.: Makromol. Chem. *179*, 949 (1978)
28) Komada, H., Okada, M., Sumitomo, H.: Polymer Preprints Jpn. *26*, 814 (1977)
29) a) Hall, Jr., H. K., DeBlauwe, Fr.: J. Am. Chem. Soc. *97*, 655 (1975)
    b) Hall, Jr., H. K., DeBlauwe, Fr., Carr, L. J., Rao, V. S., Reddy, G. S.: J. Polym. Sci.,
    Symposium No. 56, 101 (1976)
30) Okada, M., Sumitomo, H., Irii, S.: unpublished
31) Okada, M., Sumitomo, H., Komada, H.: unpublished
32) Okada, M., Sumitomo, H., Irii, S.: Makromol. Chem. *177*, 2331 (1976)
33) Kops, J., Schuerch, C.: J. Polym. Sci. Part C *11*, 119 (1965)
34) Saegusa, T., Motoi, M., Suda, H.: Macromolecules *9*, 231 (1976)
35) Kops, J., Spanggard, H.: Makromol. Chem. *176*, 299 (1975)
36) Kops, J., Spanggard, H.: Makromol. Chem. *175*, 3077 (1974)
37) Okada, M., Mita, K., Sumitomo, H.: Makromol. Chem. *176*, 859 (1975)
38) Yamashita, Y., Okada, M., Suyama, K., Kasahara, H.: Makromol. Chem. *114*, 146 (1968)
39) Hall, Jr., H. K., Schneider, A. K.: J. Am. Chem. Soc. *80*, 6409 (1958)
40) Noyce, D. S., Weingarten, H.: J. Am. Chem. Soc. *79*, 3101 (1957)
41) Collected Papers of W. H. Carothers. New York: Interscience 1940, p. 148
42) Hall, Jr., H. K., Brandt, M. K., Mason, R. M.: J. Am. Chem. Soc. *80*, 6420 (1958)
43) Hall, Jr., H. K. Banchard, Jr., E. P., Martin, E. L.: Macromolecules *4*, 142 (1971)
44) Okada, M., Sumitomo, H., Yamamoto, Y.: Makromol. Chem. *175*, 3023 (1974)
45) Okada, M., Sumitomo, H., Tajima, I.: Macromolecules *10*, 505 (1977)
46) Okada, M., Sumitomo, H., Tajima, I.: Polymer Preprints Jpn. *26*, 878 (1977)
47) Ashida, T., *et al.*: to be published
48) Goethals, E. J.: Adv. Polym. Sci. *23*, 103 (1977)
49) Okada, M., Sumitomo, H., Fujii, O.: unpublished
50) Pedersen, C. J.: J. Am. Chem. Soc. *89*, 7017 (1967)
51) Lehninger, A. L.: Biochemistry. New York: Worth Publishers, Inc. 1970, p. 614
52) Okada, M., Sumitomo, H., Tajima, I.: Polymer Preprints Jpn. *25*, 1669 (1976)
53) Moore, J. A., Kelly, J. E.: J. Polym. Sci., Polym. Lett. Ed. *13*, 333 (1975)
54) Hall, Jr., H. K.: J. Am. Chem. Soc. *82*, 1209 (1960)
55) Ogata, N., Tohoyama, S.: Bull. Chem. Soc. Jpn. *39*, 1556 (1966)
56) Ogata, N., Asahara, T., Tohoyama, S.: J. Polym. Sci., Part A-1 *4*, 1359 (1966)
57) Sumitomo, H., Hashimoto, K., Ando, M.: J. Polym. Sci., Polym. Lett. Ed. *11*, 635 (1973)
58) Sumitomo, H., Hashimoto, K.: Macromolecules *10*, 1327 (1977)
59) Sittler, E., Sebenda, J.: J. Polym. Sci., Part C *16*, 67 (1967)
60) Odian, G.: Principles of polymerization. New York: McGraw-Hill 1970, Chapt. 7
61) Sebenda, J.: J. Macromol. Sci., Chem. *6*, 1145 (1972)
62) Stehlicek, J., Labsky, J., Sebenda, J.: Collect. Czech. Chem. Commun. *32*, 545 (1967)
63) Hall, Jr., H. K.: J. Am. Chem. Soc. *80*, 6404 (1958)
64) Sorenson, W. R., Campbell, T. W.: Preparative methods of polymer chemistry. 2nd edit.
    New York: Wiley 1968
65) Overberger, C. G., Eliott, J. R., Gaylord, N. G., Bailey, W. J., Wittbecker, E. L. (eds.):
    Macromolecular syntheses. Vol. 3. New York: Wiley 1968
66) Morton, W. E., Heale, J. W. S.: Physical Properties of Textile Fibers, The Textile Institute,
    1964, p. 164
67) Sumitomo, H., Hashimoto, K., Ohyama, T.: Polymer Preprints Jpn. *26*, 1266 (1977)
68) Sumitomo, H., Hashimoto, K., Ohyama, T.: unpublished
69) Noguchi, H., Satake, I.: Polymer Preprints Jpn. *25*, 1665 (1976)
70) Sumitomo, H., Hashimoto, K.: Kobunshi Ronbunshu *34*, 747 (1977)

Received April 4, 1978
T. Saegusa (editor)

Cationic Olefin Polymerization Using Alkyl Halide Alkylaluminum Initiator Systems

1. Reactivity Studies

Joseph P. Kennedy and Prakash D. Trivedi*

Institute of polymer Science, The University of Akron, Akron, Ohio 44325, U.S.A.

After a survey of pertinent earlier studies, isobutylene polymerizations have been carried out using t-BuX initiators, Me_3Al, Et_2AlX and $EtAlCl_2$ coinitiators and MeX (X = Cl, Br, I) and n-pentane solvents in the range from $-25°$ to $-100°C$. The effect of t-BuX, Me_3Al, Et_2AlX or $EtAlCl_2$, MeX and temperature on polymerization rate has been determined. Polyisobutylene (PIB) yield, molecular weight and molecular weight distribution have been studied in detail. Initiator reactivity series have been established on the basis of polymerization rates, floor temperatures, PIB yields and initiator efficiencies. The effect of solvent on initiator reactivity order has been evaluated. These results have been explained considering details of initiation, which involves four steps: complexation, displacement, ionization and initiation. The absence of polymerization using MeI and MeBr (with Et_2AlI system) has been interpreted in terms of propagation-inactive halonium ions.

Table of Contents

*   Prakash D. Trivedi present address: Central Research Laboratories, The Firestone Tire and Rubber Company, Akron, Ohio 44317, U.S.A.

I. Introduction

The aim of this series of papers is comprehensively to survey and discuss recent results concerning the polymerization of isobutylene by a variety of alkyl halide/alkylaluminum initiator systems; in particular to review and elucidate the effects of t-BuX initiators, Me_3Al, Et_2AlX and $EtAlCl_2$ coinitiators, and MeX solvents (X = Cl, Br, I) on the rate of isobutylene polymerization, polyisobutylene (PIB) yield, molecular weight and molecular weight distribution. The effects of temperature and monomer concentration on PIB yield and molecular weight are also within the scope of this study. The first paper concerns problems of reactivities, initiator efficiency, the nature of the halogen on these parameters and on solvent effects. The subsequent contribution[1] will focus on molecular weights, in particular on the effect of reaction conditions on molecular weights and molecular weight distributions.

Impetus for these studies arose as a consequence of steadily increasing commercial and scientific interest in alkylaluminum compounds in general and in their polymerization initiating ability in particular. While since about two decades alkylaluminums have acquired important roles as essential components of Ziegler-Natta coordination catalysts, the discoveries that these compounds are also efficient coinitiators for a large variety of cationic polymerizations, block and graft copolymerizations, and derivatizations are of much later vintage. The key discovery for the latter developments was the controlled cationic olefin polymerization, that initiation of isobutylene and other cationically polymerizable monomers can be readily controlled by the use of certain alkylaluminum compounds $e.g.$, Et_2AlCl, Et_3Al[2]. Until this discovery initiation of cationic polymerizations was a largely irreproducible, chancy undertaking which relied on the (fortunately ubiquitous) availability of protogenic impurities $e.g.$, moisture. It is well known that conventional Lewis acids like $AlCl_3$, $TiCl_4$, $SnCl_4$, $ZrCl_4$ or BF_3 when added to reactive polymerizable alkenes immediately induce sometimes even explosive polymerizations. In contrast, certain alkylaluminums of somewhat weaker Lewis acidities than the above conventional Friedel-Crafts halides do not alone initiate polymerizations but require the *purposeful* addition of a cationogen[2].

Further, while conventional Friedel-Crafts halides produce high molecular weight polyisobutylenes or polyisobutylene copolymers ($e.g.$, butyl rubbers, IIR) only at relatively low ($\sim -100\,^\circ$C) temperatures, alkylaluminum-based initiator systems produce high molecular weight materials at much higher ($\sim -40\,^\circ$C) temperatures. In this series of publications a comprehensive theory will be developed to explain these differences.

Since initiation with conventional Friedel-Crafts halides cannot be controlled, the fine-tuning of reactions becomes extremely cumbersome. In contrast, by the use of alkylaluminum compounds elementary events (initiation, termination, transfer) become controllable and thus molecular engineering becomes possible. Indeed, by elucidating the mechanism of initiation etc., a large variety of new materials, $i.e.$, block[3], graft[4-6] bigraft[7] copolymers, have been synthesized and some of their physical-chemical properties determined.

Kennedy[8, 9] studied in detail aspects of initiation of isobutylene and styrene polymerization using Et_2AlCl coinitiator. Subsequently, Kennedy and co-workers[10] studied the mechanism of initiation and termination by model reactions. Using Me₃Al and t-BuX, these authors calculated rate constants for neopentane formation using MeX solvents. These studies provided an understanding of the effect of halogen in t-BuX and MeX on initiation. Experiments with 2,4,4-trimethyl-1-pentene, a nonpolymerizable monomer, provided valuable insight into the mechanism of isobutylene polymerization with alkylaluminum systems, in particular to an increased understanding of the mechanism of initiation, transfer and termination[11].

The first paper of this series concerns the effects of t-BuX, Me₃Al, Et₂AlX and EtAlCl₂, and MeX on PIB yields and polymerization rates. The second paper[1] will survey and discuss the effects of reaction variables on molecular weights of PIB and molecular weight control in isobutylene polymerization.

II. Historical Background

Although the history of isobutylene polymerization chemistry can be traced back to the nineteenth century[12], systematic research began barely forty years ago. The introduction of alkylaluminum coinitiators took place at about the same time[13]. Since early work has been reviewed by Kennedy and Gillham[14], only a comprehensive review of recent study of isobutylene polymerization using alkylaluminum coinitiators will be presented.

Kennedy and co-workers[10] studied the kinetics of the reaction between Me₃Al and t-butyl halides using methyl halide solvents as a model for initiation and termination in cationic polymerization. Neopentane was generated rapidly, without side reactions and rates were determined by NMR spectroscopy. The major conclusions were:

1) Rates of alkylation decreased as: t-BuCl > t-BuBr > t-BuI.

2) For a given t-butyl halide, the rate was dependent on solvent as: MeCl > MeBr > MeI ≫ cyclopentane.

3) The overall activation energies using methyl halide and cyclopentane were 11 kcal/mole and 16 kcal/mole, respectively.

4) The overall rate was a function of steric requirement and/or basicity of the halogen in t-BuX.

The apparent second order rate constants[10] of t-BuX with Me₃Al at −40 °C are given in Table 1. The proposed mechanism to explain the above results was as follows. A half-bridged intermediate is assumed to be formed in the presence of MeCl.

Table 1. Rates of alkylation of t-butyl halides with trimethylaluminum at $-40\,^\circ\text{C}$[10]

| t-BuX | k, l/mole-sec $\times 10^6$ | | | |
	MeCl	MeBr	MeI	C_5H_{10}
t-BuCl	$> 10^{9)a}$	4100	2000	1.2
t-BuBr	13,000	1800	100	0.7^b
t-BuI	75^c	15^c	3.2^c	0.005^c

a   Calculated from the observed rate at $-80\,^\circ\text{C}$, assuming an activation energy of 10 kcal/mole.
b   Calculated from the measured rates in the temperature range 0 to $+50\,^\circ\text{C}$.
c   Calculated from the measured rates in the temperature range -20° to $+20\,^\circ\text{C}$.

The half-opened dimer was assumed to further dissociate into monomeric complex, $Me_3Al{\leftarrow}MeCl$. Subsequently t-BuCl replaced MeCl giving $Me_3Al{\leftarrow}t$-BuCl which ionized and finally collapsed giving neopentane. The enormous difference in the rates of neopentane formation was attributed to the differences in basicity and steric requirements of the halogen. t-BuCl was most active, since $-Cl$ is the most basic and smallest of the halogens used. The lowest rate with t-BuI was due to the low basicity and unfavorable steric compression of the $-I$ substituent. The difference in rates in regard to the MeX solvents used was proposed to be due to the difference in polarizability of the halogens. The less polarizable $-Cl$ formed the weakest $Me_3Al{\leftarrow}MeCl$ coordinate bond, while the more polarizable $-I$ formed the strongest coordinate bond. As the rate of MeX displacement was governed by the strength of the coordinate bond, it was highest for MeCl and lowest for MeI. It is also possible that the rate of alkylation of t-Bu cations is further reduced in MeBr and MeI due to the formation of halonium ions $Me_3C{-}\overset{\oplus}{Br}{-}Me$ and $Me_3C{-}\overset{\oplus}{I}{-}Me$, respectively (Section VIII).

Recently, Kennedy and Rengachary[11] studied cationic olefin model and polymerization reactions. Important conclusions of the model study were:

1) Coinitiator reactivity decreased as $Me_2AlCl > Et_2AlCl > Me_3Al > Et_3Al$.

2) The counterion strongly affected the relative rates of elimination vs. alkylation (transfer vs. termination).

3) The sequence of termination rates based on conversions was: $Et_3Al > Me_3Al > Et_2AlCl > Me_2AlCl$.

4) Alkylaluminum containing a β-hydrogen relative to aluminum reacted preferentially by hydridation (reduction).

Important conclusions regarding the polymerization study were:

1) t-BuCl/Me_2AlCl produced highest molecular weight PIB in the range from -50° to $-70\,^\circ\text{C}$.

2) Stable counterions like $Me_2AlCl_2^\ominus$ and $Me_2AlClBr^\ominus$ yielded higher molecular weight PIB than less stable counterions like $Et_2AlCl_2^\ominus$ and $Et_3AlCl^\ominus$.

3) $\Delta E_{\bar{M}_v}$ for t-BuCl/Et_3Al and t-BuCl/Et_2AlCl systems was ~ -2 kcal/mole, for t-BuBr/Et_3Al and t-BuBr/Et_2AlCl systems was ~ -4 kcal/mole and for $MeAlCl_2$, t-BuCl/Me_2AlCl and t-BuBr/Me_2AlCl systems was ~ -7 kcal/mole.

4) Initiator efficiency in terms of conversions and molecular weights were similar for model compounds and polymerizations. The influence of chlorine and bromine-containing counterions on polymerization was similar to that found in model study.

A subsequent paper[15] in this series dealt with initiation with electrophilic halogens. The important findings were as follows:

1) Both chlorine and bromine in conjuction with alkylaluminums were active initiators in isobutylene polymerization.

2) Initiator efficiency in isobutylene and styrene polymerization decreased as, $Cl_2 > Br_2 \gg I_2 = 0$. I_2 did not initiate.

3) The molecular weights of PIB obtained with Cl_2/Et_2AlCl and t-BuCl/Et$_2$AlCl systems were the same, as also, the molecular weights of PIB prepared using Cl_2/Me_3Al and t-BuCl/Me$_3$Al systems suggesting that they were controlled by the counterions.

Marek and co-workers[16] studied the polymerization of isobutylene using $EtAlCl_2$ and n-heptane. They found:

1) The rate of polymerization in the range from $+21°$ to -55 °C was constant, *i.e.* the activation energy was zero.

2) The initial rate was first order in monomer and second order in $EtAlCl_2$.

3) The relative rate constants found were $k_{tr}/k_p = 2.8 \times 10^{-2}$ and $k_t/k_p = 4.3 \times 10^{-3}$ at -10 °C.

4) $\Delta E_{\bar{M}_v}$ was -5.8 kcal/mole in the range from $+21°$ to -55 °C.

5) Extraneously added water was found to be an inhibitor. The authors suggested that initiation might occur by the reaction of monomer with the ion-pair generated either directly from the dimer $(EtAlCl_2)_2$ or by a reaction of two $EtAlCl_2$ molecules. The evidence presented, however, was insufficient to establish a satisfactory mechanism.

In a series of patents[17−20], Italian workers disclosed copolymerization of isobutylene and isoprene using Et_2AlCl and a host of initiators, *e.g.* R′—C—X, where
$$\overset{\|}{Y}$$
X = halogen, R′ = alkyl group, Y = sulfur or nitrogen; YZR′X, X = halogen, Z = nitrogen, carbon or phosphorus, R′ = alkyl group and Y = organic group like $-SO_2R$. Though these patents open up many interesting possibilities, the inventors have not reported endgroup characterization or tried to elucidate the nature of these initiators.

In a series of seven recent publications, these Italian authors[21−27] reported isobutylene homo- and copolymerizations using alkylaluminum coinitiators in the presence of halogen, interhalogen compounds and alkyl halide initiators. The following conclusions[21] are reported in the first paper.

1) Initiator efficiencies decreased as $Cl_2 > ICl > IBr > Br_2 > I_2$.

2) Highest molecular weights were obtained with Cl_2 and Br_2, while molecular weights were low and independent of the nature of initiators ICl, IBr and I_2.

3) Br_2 and I_2 did not initiate in conjunction with Et_3Al.

4) Whereas I_2 was inert with Et_3Al and inefficient in conjunction with Et_2AlCl and Et_2AlBr, it was an efficient initiator when used with Et_2AlI.

The second paper[22] dealt with model reactions and showed that many side reactions occurred between halogens and the other reagents used and experiments had to be designed to minimize these reactions. Using 2,4,4-trimethyl-1-pentene, the authors showed that initiation was due to electrophilic halogen.

The third paper[23] reported a study with the I_2/Et_2AlI initiator system. Though the I_2/Et_2AlI system was not efficient in n-pentane, high initiator efficiencies were obtained using EtCl or CH_2Cl_2. The molecular weight of isobutylene-isoprene co-polymers increased as the temperature decreased. Using CH_2Cl_2, $\Delta E_{\overline{M}_v}$ was -2.87 kcal/mole.

Subsequent two papers[24, 25] of this series dealt with isobutylene homo- and copolymerization using the Cl_2/Et_2AlCl initiator system. Isobutylene polymeriza-tion was more reproducible using MeCl than CH_2Cl_2. Maximum polymerization rate varied linearly with $[Et_2AlCl]$ and $[Cl_2]$. $\overline{DP}_v$ decreased with the increase in $[Et_2AlCl]$ and $[Cl_2]$ and the maximum rate of polymerization increased linearly with $[i-C_4H_8]$. $\overline{M}_v$ increased with increasing monomer concentration and the Mayo plot ($1/\overline{DP}_v$ vs. $1/[M]$) gave a straight line passing through the origin, indicating negligible monomer transfer. The maximum rate of polymerization decreased and $\overline{M}_v$s increased with a decrease in temperature. Arrhenius plots gave $\Delta E_{\overline{M}_v} = -2.87$ kcal/mole and $\Delta E_{Rate} = +4.75$ kcal/mole. $\Delta E_{\overline{M}_v}$ decreased with increase in isoprene concentration in copolymerizations. Addition of water seemed to increase conversion when Et_2AlCl was used. The physical properties of butyl rubbers pre-pared using Cl_2/Et_2AlCl at $-40\ °C$ and that of a commercial sample were similar.

The final two papers of these Italian investigators[26, 27] concern a comparative study of alkylaluminum-based initiators with those of conventional Friedel-Crafts acids like $AlCl_3$ and $EtAlCl_2$. It was reported that in conjunction with Et_2AlCl coinitiator, Cl_2 is a better initiator than either HCl or t-BuCl. The t-BuCl/Et_2AlCl initiator system is characterized by slow initiation and absence of monomer trans-fer, however important termination occurs. In contrast, with the $AlCl_3$ and $EtAlCl_2$ systems, initiation is very fast and at least one important termination re-sulting in incomplete conversion takes place.

III. Experimental

1. Materials

Isobutylene, methyl chloride, methyl bromide, ethyl chloride (Linde Div. Union Carbide Corp.) were obtained in high purity and were further purified by passing through a column containing barium oxide and molecular sieves.

2,4,4-Trimethyl-1-pentene, methyl iodide, t-butyl chloride, t-butyl bromide (Matheson Coleman and Co.) and t-butyl iodide (Eastman Kodak) were obtained in highest purity and were distilled over calcium hydride or molecular sieves and stored at Dry-Ice temperature.

n-Pentane (Matheson Coleman and Co.) was dried over calcium hydride, distilled and then stirred for 24 hours with small amounts of Et_3Al and then redistilled.

The alkylaluminums were obtained from Texas Alkyl Inc. and were purified by vacuum distillation under nitrogen atmosphere. B. P. °C/mmHg: Me_3Al: 60°/70; Et_2AlCl: 125°/50; Et_2AlBr: ~150°/50 and $EtAlCl_2$: 100°/30. Et_2AlI was used as received. Prior to distillation, Et_2AlCl and Et_2AlBr were stirred for one hour over sodium chloride and sodium bromide, respectively, at 80 °C and stored over these salts. All alkylaluminums were stored at Dry-Ice temperature.

2. Polymerization Studies

Polymerizations were carried out in Pyrex test tubes in a stainless steel enclosure equipped with a cooling bath, inlets for gaseous reagents and a water analyzer. Nitrogen gas was flushed continuously to maintain a moisture level below 30 ppm.

The gaseous reagents were distilled and collected inside the enclosure. The 10 vol.% alkylaluminum and 1 vol.% t-butyl halide solutions were freshly prepared. The reaction mixtures were stored manually or by a vortex stirrer. Since PIB tended to precipitate out of solution in highly swollen form, relative rates could thus be established by visual observation.

Temperatures ranged from −20 °C to −90 °C. Reactions were terminated after 20 to 60 minutes by adding 5 ml prechilled MeOH. The quenched mixture was transferred to a preweighed aluminum dish and dried in a vacuum oven at 50 °C.

The effect of monomer concentration was studied using n-pentane solvent and maintaining the total volume of isobutylene plus n-pentane constant. Methyl halide concentration was kept constant so as to maintain constant medium polarity. Attempts were made to keep conversions below 20%. At −30 °C, due to almost explosive polymerizations, conversions could only be maintained below 40%.

2,4,4-Trimethyl-1-pentene (TMP) (2 ml) was mixed with MeX (5 ml) and desired amounts of alkylaluminum and t-BuX solutions were added. In a similar study, 2 ml TMP were mixed with 5 ml MeBr and 0.1 ml of a solution of $AlBr_3 \cdot MeBr$ complex in MeBr (10 vol.%) was added at −80 °C. The reaction was stopped after 20 mins. and the products separated and analyzed using Kennedy and Rengachary's method[11].

3. Molecular Weight Determination

Attempts were made to determine number average molecular weights ($\overline{M}_n$) by osmometry (Mechrolab Model 502, high speed membrane osmometer, 1 to 10 g/l toluene solution at 37 °C), however, in many instances irreproducible data were obtained, probably due to the diffusion of low molecular weight polymer through the membrane. This technique was abandoned in favor of gel permeation chromatography (GPC).

GPC was carried out using a Waters Associates' Model 100 instrument and 0.4% THF solutions of PIB. The GPC unit was equipped with seven polystyrene gel columns having pore sizes (in Å): $7 \times 10^5 - 5 \times 10^6$, $1.5 \times 10^5 - 7 \times 10^5$, $5 \times 10^4 - 1.5 \times 10^5$, $1.5 \times 10^4 - 5 \times 10^4$, $5 \times 10^3 - 1.5 \times 10^4$, $5 \times 10^3 - 1.5 \times 10^4$ and 2000−5000. The calibration curve was prepared using fractionated PIB samples of narrow molecular weight distributions $M_w/M_n (\sim 1.5)$. Light scattering was used to determine weight average molecular weights $\bar{M}_w$ for samples used for the calibration curve. The molecular weights were calculated using the formula:

$$\bar{M}_n = \frac{\Sigma P_i}{\Sigma P_i/M_i} \quad \text{and} \quad \bar{M}_w = \frac{\Sigma P_i M_i}{\Sigma P_i}$$

where P_i is the height of a point on the GPC trace at the ith elution count and M_i is molecular weight corresponding to that elution count obtained from the calibration curve. It was not possible to run duplicates with all samples, hence only spot checks were made. Reproducibility was about 15%. The GPC curves shown are normalized. Refractive index measurements were used to determine polymer concentration.

Viscosity average molecular weights $\bar{M}_v$ were determined using a Ubbelohde viscometer and diisobutylene solutions at 20 °C with at least three dilutions for every solution. The $\bar{M}_v$ was calculated from intrinsic viscosity[28]. Averages of two determinations are reported. Reproducibility was ~10%.

The activation energy differences of $\bar{M}_v$ as well as of $\bar{M}_n$ and $\bar{M}_w$; and k_{tr}/k_p and k_t/k_p were calculated from Arrhenius and Mayo plots, respectively, by linear regression analysis using a computer. The $\Delta E_{\bar{M}_w}$ values given in kcal/mole can be converted to kJ/mole by multiplying with 4.18.

IV. A Note on Terminology

The terminology proposed by Kennedy[29] is used according to which the source of the initiating entity, *i.e.* the proton or carbocation, is termed the initiator and the Lewis acid or Friedel-Crafts acid which aids the generation of cation is the coinitiator. This terminology was extended to embrace systems in which the *de facto* initiator was advantitious traces of protogenic impurities and not the Friedel-Crafts acids ostensibly used for that purpose. It is proposed to use for these systems quotation marks around the suspected initiator, ubiquitous water, *e.g.*, "H_2O"/BF_3. The quotation mark is used to express the requirement for an initiator which, however, was not purposefully introduced. The "H_2O" also symbolizes unidentified impurities, though the overwhelming majority of investigations indicate H_2O to be the initiating entity. Only in instances when the initiator was added purposefully to initiate polymerization, *i.e.*, in rigorously controlled stopping experiments, should the initiator be written without quotation mark *e.g.*, H_2O/BF_3.

For systems which have been established by rigorous high vacuum experiments not to require the separate addition of cationogen, the use of Friedel-Crafts acid alone is recommended, *i.e.* EtAlCl$_2$/i-C$_4$H$_8$ [16], TiCl$_4$/i-C$_4$H$_8$ [30].

Finally, it is recommended to refer to polymerization systems in terms of four components; initiator/coinitiator/monomer/solvent, *e.g.* t-BuCl/Et$_2$AlCl/i-C$_4$H$_8$/MeCl. In cases where the identity of monomer or solvent is obvious, initiator/coinitiator sufficiently define the system. The symbol for counteranion, $G^\ominus$, is independent of initiator and coinitiator used. For example, $Et_2AlClBr^{\ominus}$ could arise either from Et_2AlCl + t-BuBr or Et_2AlBr + t-BuCl.

V. Isobutylene Polymerization Using t-BuX/Me$_3$Al Initiator Systems and MeX Solvents

1. Introduction

Kennedy and co-workers[10] studied model cationic polymerization initiation and termination. They determined the effect of halogens in t-BuX and MeX on the rate of reaction between t-BuX and Me$_3$Al. The pseudo second order rate constant decreased (Table 1) as:

t-BuCl > t-BuBr > t-BuI and MeCl > MeBr > MeI > cyclopentane.

It was postulated that the rate decreased as the basicity of the halogen decreased and/or steric compression increased in t-BuX, and as the polarizability of halogen in MeX increased. The objective of the present research was to extend this model study to isobutylene polymerization systems, in particular to investigate the effect of reagent addition sequence and that of the nature of the halogen in t-BuX and MeX on the polymerization rate and PIB yield using Me$_3$Al coinitiator.

2. Effect of Reagent Addition Sequence on Isobutylene Polymerization

The objective of this phase of investigations was to study the effect of reagent introduction sequence on PIB yield and overall polymerization rate using t-BuX/Me$_3$Al/MeCl, where X = Cl, Br at $-40°$ and -50 °C. With the t-BuCl/Me$_3$Al/MeCl system, polymerization rates and PIB yields at $-40°$ and $-50°$ were independent of reagent addition sequence. Thus, addition of either t-BuCl or Me$_3$Al to quiescent Me$_3$Al/i-C$_4$H$_8$/MeX or t-BuCl/i-C$_4$H$_8$/MeX mixtures produced high, sometimes explosive, polymerization rates and yields.

The results obtained with t-BuBr initiator are given in Table 2. At -40 °C, rates were explosive only when the reagent addition sequence was i-C$_4$H$_8$/t-BuBr/Me$_3$Al, where slow rates were obtained with i-C$_4$H$_8$/Me$_3$Al/t-BuBr. At -50 °C, high rate and yield were obtained only when the sequence was i-C$_4$H$_8$/t-BuBr/Me$_3$Al and the Me$_3$Al was not cooled to -50 °C. Negligible yields and slow rates were observed with the same reagent addition sequence but cooling the Me$_3$Al to -50 °C or by changing the sequence to i-C$_4$H$_8$/Me$_3$Al/t-BuBr.

These results may be explained assuming complexation between Me$_3$Al and i-C$_4$H$_8$. Due to the relatively high basicity of t-BuCl, reaction between t-BuCl and Me$_3$Al is fast even in the presence of Me$_3$Al · i-C$_4$H$_8$ complexes. Complexation, however, reduces the rate and yield with the less basic t-BuBr. Thus, adding Me$_3$Al to i-C$_4$H$_8$ followed by t-BuBr addition produces slow rates at -40 °C and significantly inhibits polymerization at -50 °C. It seems that t-BuBr reacts slowly with the complex at -40 °C, while it is unable to break the complex at -50 °C. While

Table 2. Effect of reagent addition sequence on isobutylene polymerization

Time °C	Time min.	Yield[a] %	Addition sequence	Comments
−40°	60	87 ± 2	i-C$_4$H$_8$/Me$_3$Al/t-BuBr	Slow polymerization
−40°	5	85 ± 2	i-C$_4$H$_8$/t-BuBr/Me$_3$Al[b]	Fast polymerization
−50°	5	90 ± 2	i-C$_4$H$_8$/t-BuBr/Me$_3$Al[b]	Fast polymerization
−50°	60	5 ± 2	i-C$_4$H$_8$/t-BuBr/Me$_3$Al[c]	Slow polymerization
−50°	60	3 ± 2	i-C$_4$H$_8$/Me$_3$Al/t-BuBr	Slow polymerization

[Isobutylene] = 3.0 M.
[Me$_3$Al] = 2.5 × 10^{-2} M.
[t-BuBr] = 8.9 × 10^{-3} M.
MeCl = 15 ml.

[a] Yield: Average of two runs.
[b] Me$_3$Al was added without cooling to reaction temperature.
[c] Me$_3$Al was cooled to −50 °C before adding.

direct spectroscopic evidence for complexation between Me$_3$Al and isobutylene is absent, kinetic evidence has been presented by Kennedy and co-workers[31, 32].

In sum, while the reagent addition sequence is unimportant with t-BuCl, it strongly affects the rate and yield using t-BuBr. This effect is attributed to complexation between Me$_3$Al and i-C$_4$H$_8$. Since rates were readily controllable by the i-C$_4$H$_8$/Me$_3$Al/t-BuBr addition sequence, this reagent addition sequence was used in subsequent experiments.

3. Effect of t-BuX Initiator, MeX Solvent and Temperature on Isobutylene Polymerization

The effect of t-BuX initiator and MeX solvent on PIB yield was studied in the temperature range from −20° to −60 °C. Results are shown in Table 3. Initial rates of polymerization at −40 °C were determined from time-conversion plots (Fig. 1).

According to the data in Table 3, the t-BuCl/Me$_3$Al/MeCl system produces higher polymer yield and is active over a wider temperature range than any other initiator/solvent combinations. While yields obtained with t-BuCl and t-BuBr using MeCl are comparable up to about −40 °C, lower yields are obtained with t-BuBr than with t-BuCl below −40 °C.

For a given initiator, lower yields are obtained using MeBr than MeCl. Initiator efficiency (monomer conversion in M/initiator concentration in M) calculated for data obtained at −40 °C are given in Table 8. While initiator efficiencies were nearly the same for t-BuCl and t-BuBr using MeCl, t-BuCl was more efficient than t-BuBr using MeBr. Polymerization did not occur with t-BuI or using MeI solvent.

The effect of temperature is similar in these systems. Thus, yield was little affected or was reduced with decrease in temperature. At a certain temperature level, depending on the nature and concentration of initiator and solvent, yields dropped to zero. This floor temperature is −60 °C for t-BuCl/MeCl and t-BuBr/MeCl and −40 °C for t-BuCl/MeBr and t-BuBr/MeBr.

The initial rate of polymerization was determined from the initial slopes of time-conversion curves (Fig. 1) using t-BuX/Me$_3$Al/MeCl systems at −40 °C. This

Table 3. Effect of temperature on yield of PIB prepared using t-BuX/Me$_3$Al/MeX system

Temp. °C	[t-BuX] M × 10³	Polyisobutylene Yield %					
		t-BuCl		t-BuBr		t-BuI	
		MeCl	MeBr	MeCl	MeBr	EtCl	MeI
0°	50.0	–	–	–	–	0	0
−20°	8.9	–	70.5	–	73.5	–	–
	20.0	–	–	–	–	0	0
	50.0	–	–	–	–	0	0
−25°	4.5	–	30.8	–	26.0	–	–
	8.9	–	71.7	–	49.2	–	–
−30°	4.5	72.1	23.3	83.5	–	–	–
	8.9	74.0	72.0	86.5	12.2	–	–
−35°	4.5	73.2	–	63.3	0	–	–
	8.9	72.0	53.0	87.4	2.5	–	–
−40°	4.5	65.0	–	83.0	–	0	0
	8.9	81.8	49.0	84.0	0	0	–
−45°	8.9	84.0	0	14.7	–	–	–
−50°	4.5	24.3	–	0	–	–	–
	8.9	90.2	–	14.3	–	–	–
	45.0	–	–	89.3	–	–	–
−60°	4.5	0	–	0	–	–	–
	8.9	0	–	0	–	–	–

[Isobutylene] = 3.0 M, [Me$_3$Al] = 2.5 × 10^{-2} M.
MeX = 15 ml, EtCl = 15 ml, Time 60 mins.

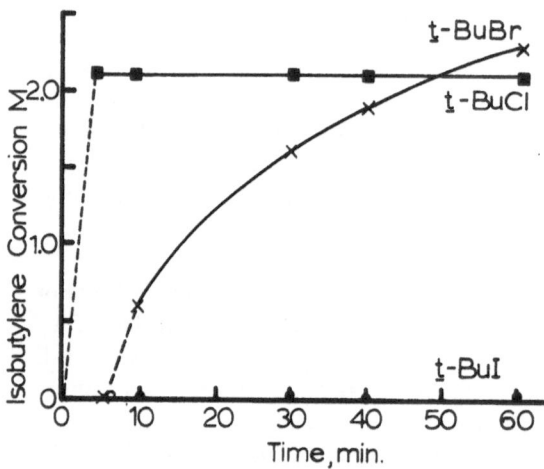

Fig. 1. Rate of isobutylene polymerization using t-BuX/Me$_3$Al/MeCl ([Isobutylene] = 3.0 M, [Me$_3$Al] = 2.5 × 10^{-2}M, [t-BuX] = 4.5 × 10^{-3}M, MeCl = 15 ml. −40 °C)

temperature was chosen since at −30 °C the rates were too high while at −50 °C t-BuBr did not initiate polymerization. The rates obtained were 8.3 × 10^{-3}, 1.7 × 10^{-3} and 0 mole/l.sec using t-BuCl, t-BuBr and t-BuI, respectively.

t-BuI did not initiate polymerization of isobutylene or dimerization of 2,4,4,-trimethyl-1-pentene. Thus, polymerization did not occur even at relatively high

t-BuI concentration (5×10^{-2} M) or temperature (0 °C using EtCl). Similarly, t-BuI/Me$_3$Al (1.6×10^{-3} and 6.3×10^{-3} M, respectively) did not t-butylate 2,4,4-trimethyl-1-pentene (6.3×10^{-3} M) at 0 °C using EtCl. Evidently, t-BuI is not an initiator in conjunction with Me$_3$Al.

Polymerization did not occur using MeI solvent even at high t-BuCl or t-BuBr concentrations and temperatures. Polymerization occurred, however, using mixtures of MeCl and MeI. Details of this study are given in Section VII.

Combining all these findings, $i.e.$ initiator efficiencies, polymerization rates and yields, and floor temperatures, a relative order of initiator reactivities can be obtained. For the t-BuX/Me$_3$Al/MeX systems, the initiator reactivity is: t-BuCl > t-BuBr $\gg$ t-BuI = 0. The nature of solvent also affects initiator reactivity as follows: MeCl > MeBr $\gg$ MeI = 0.

Yields are determined by comparative rates of initiation, propagation, termination and possibly transfer, (R_i, R_p, R_t and R_{tr}, respectively). Assuming R_p and R_{tr} to be similar for all systems, high yields would imply high R_i and/or low R_t. Kennedy and co-workers[10] found in their model cationic polymerization, initiation and termination rate trends similar to the above initiator reactivity orders and attributed their findings to differences in R_i. The similarity between the results of model and polymerization studies indicate that R_i is more important than R_t in polymerizations also, and that the higher overall rates and yields as well as higher initiator efficiencies obtained with t-BuCl are due to relatively high R_i. Since t-BuI gave the slowest rates in model study[10], it was not surprising that t-BuI did not initiate isobutylene polymerization.

In line with the conclusions derived in the model study[10], the effect of solvent on initiator reactivity can also be explained on the basis of halogen polarizability in MeX. A detailed discussion of these trends is given in Section VIII. MeI is a poison in isobutylene polymerization. The poisoning activity of MeI will be discussed in Section VII.

In sum, kinetic evidence for the existence of Me$_3$Al $\cdot$ i-C$_4$H$_8$ complex is given by the reagent addition sequence study. Initiator reactivity trends of t-BuX/Me$_3$Al/MeX systems in model experiments[10] and polymerizations are similar. Initiator reactivity decreases as t-BuCl > t-BuBr $\gg$ t-BuI = 0 and is affected by solvent as: MeCl > MeBr $\gg$ MeI = 0. The similarity of the trends in model[10] and polymerization studies strongly suggests that the overall polymerization rate and yield are determined by R_i in t-BuX/Me$_3$Al/MeX systems. Incomplete conversions and higher yields at increased initiator concentrations indicate kinetic termination in these systems.

VI. Isobutylene Polymerization Using t-BuX/Et$_2$AlX and "H$_2$O"/ EtAlCl$_2$ Initiator Systems and MeX Solvents

1. Introduction

The objective of this phase of investigation was to study the details of isobutylene polymerization using t-BuCl, t-BuBr and t-BuI initiators, Et$_2$AlCl, Et$_2$AlBr and

Et_2AlI coinitiators and MeCl, MeBr and MeI solvents at various temperatures. The "H_2O"/$EtAlCl_2$/n-pentane system was also briefly investigated. A large number of comparative molecular weight and conversion data were gathered. The effect of MeX on the polymerization was investigated in detail. In this section, conversion and initiator efficiency data will be discussed.

2. Effect of t-BuX, MeX and Temperature on Isobutylene Polymerization Using Et_2AlCl Coinitiator

PIB yields and molecular weights ($\overline{M}_v$s) obtained using Et_2AlCl in conjunction with t-BuCl and t-BuBr initiators and MeCl and MeBr solvents are given in Table 4. Initiator efficiencies determined at $-60\,°C$ are reported in Table 8. Polymerization was absent using t-BuI initiator or MeI solvent.

t-BuCl was slightly more efficient than t-BuBr using MeCl only below $-60\,°C$. However, it was markedly more efficient than t-BuBr using MeBr. Polymerization was absent using t-BuBr below $-45\,°C$, while t-BuCl was active down to $-70\,°C$ when MeBr was used as a solvent. Also, rates appeared to be consistently higher with t-BuCl than with t-BuBr.

Yields were higher using MeCl than MeBr. Also, the floor temperature was higher using MeBr than MeCl for both t-BuCl and t-BuBr initiators.

The effect of temperature on PIB yield was dependent on the nature and concentration of t-BuX and on solvent. For t-BuX/MeCl, yields were essentially unchanged in the range from $-30°$ to $-70°C$. Except for the lowest t-BuCl concentration ($4.5 \times 10^{-4}M$), yield decreased slightly at and below $-60\,°C$. For t-BuCl/MeBr, the temperature effect was insignificant from $-25\,°C$ to $-55\,°C$. While the yield decreased at and below $-60\,°C$, it was zero at $-65\,°C$ at the lowest t-BuCl concentration (1.9×10^{-4} M). For t-BuBr/MeCl yield was not affected in the range from $-35°$ to $-45\,°C$ but decreased at $-30°C$ and below $-45\,°C$. For the t-BuBr/MeBr system, yields increased with decrease in temperature from $-25°$ to $-40\,°C$, decreased at $-45\,°C$, and polymerization was absent below $-45\,°C$. None of these systems yielded polymer at or below $-75\,°C$.

Initiator efficiencies obtained with [t-BuX] = 1.9×10^{-4} M at $-60\,°C$ (See Table 8) decrease as t-BuCl/Et_2AlCl/MeCl > t-BuBr/Et_2AlCl/MeCl > t-BuCl/Et_2AlCl/MeBr $\gg$ t-BuBr/Et_2AlCl/MeBr, t-BuI/Et_2AlCl/MeCl = 0.

Based on initiator efficiencies, polymerization rates and floor temperatures, relative initiator reactivity in conjunction with Et_2AlCl was: t-BuCl > t-BuBr $\gg$ t-BuI = 0. Further, depending on the nature of solvent, initiator reactivity decreased as MeCl > MeBr $\gg$ MeI = 0.

3. Effect of t-BuX, MeX and Temperature on Isobutylene Polymerization Using Et_2AlBr Coinitiator

Though Et_2AlBr coinitiator for isobutylene polymerization has been mentioned in the patent literature[33], detailed work with this coinitiator has not been reported. Table 5 lists PIB yields and molecular weights obtained using the t-BuX/Et_2AlBr/MeX

Table 4. Effect of temperature on yield and $\bar{M}_v$ of PIB prepared with t-BuX/Et$_2$AlCl/MeX systems

Temp. °C	t-BuX $\times 10^{-4}$ M	t-BuCl/MeCl Yield %	$\bar{M}_v \times 10^{-3}$	t-BuCl/MeBr Yield %	$\bar{M}_v \times 10^{-3}$	t-BuBr/MeCl Yield %	$\bar{M}_v \times 10^{-3}$	t-BuBr/MeBr Yield %	$\bar{M}_v \times 10^{-3}$
−25°	1.9	—	—	35.1	—	—	—	25.0	—
	3.7	—	—	51.0	—	—	—	45.0	—
−30°	0.45	19.2	—	—	—	19.5	—	—	—
	1.9	58.4	—	32.5	—	27.7	166.0	29.7	—
	3.7	—	—	50.7	155.0	—	—	50.0	211.5
−35°	0.45	26.6	—	—	—	20.6	—	—	—
	1.9	68.7	180.0	36.3	—	68.7	280.0	35.3	—
	3.7	—	—	55.1	182.0	—	—	58.8	246.7
−40°	0.45	20.6	—	—	—	13.4	—	—	—
	1.9	69.0	492.8	36.8	—	77.0	341.0	20.4	—
	3.7	—	—	52.4	239.0	—	—	62.3	271.0
−45°	0.45	17.4	—	—	—	20.7	—	—	—
	1.9	38.6	—	27.7	—	64.4	683.0	14.6	—
	3.7	76.8	—	56.9	330.0	78.4	—	37.5	385.0
−50°	0.45	13.8	—	—	—	2.4	—	—	—
	1.9	68.7	549.0	29.5	—	46.0	549.0	—	—
	3.7	—	—	63.6	345.0	—	—	—	—
−55°	0.45	17.0	—	—	—	17.3	—	—	—
	1.9	41.2	625.0	37.9	—	52.2	635.0	—	—
	3.7	80.8	—	60.0	330.0	87.3	—	—	—
−60°	0.45	6.6	—	—	—	6.4	—	—	—
	1.9	58.1	—	7.5	—	17.0	—	—	—
	3.7	—	—	30.9	389.5	—	—	—	—

Table 4. (continued)

Temp. °C	t-BuX × 10^{-4} M	t-BuCl/MeCl Yield %	$\bar{M}_v$ × 10^{-3}	t-BuCl/MeBr Yield %	$\bar{M}_v$ × 10^{-3}	t-BuBr/MeCl Yield %	$\bar{M}_v$ × 10^{-3}	t-BuBr/MeBr Yield %	$\bar{M}_v$ × 10^{-3}
−65°	0.45	8.5	—	—	—	—	—	—	—
	1.9	62.6	—	0	—	—	—	—	—
	3.7	87.7	—	40.0	—	—	—	—	—
−70°	0.45	7.3	—	—	—	4.7	—	—	—
	1.9	52.0	736.0	0	—	13.1	736.0	—	—
	3.7	—	—	37.4	460.6	—	—	—	—

[Isobutylene] = 3.0M, [Et$_2$AlCl] = 6.0 × 10^{-3} M, MeX = 15 mL, 20−60 min.

Table 5. Effect of temperature on the yield and $\bar{M}_v$ of PIB prepared with t-BuX/Et$_2$AlBr/MeX systems

Temp. °C	t-BuX × 10^4 M	t-BuCl/MeCl		t-BuCl/MeBr		t-BuBr/MeCl		t-BuBr/MeBr		t-BuI/MeCl
		Yield %	$\bar{M}_v$ × 10^{-3}	Yield %	$\bar{M}_v$ × 10^{-3}	Yield %	$\bar{M}_v$ × 10^{-3}	Yield %	$\bar{M}_v$ × 10^{-3}	Yield %
−25°	1.2	—	—	19.2	—	—	—	16.2	—	—
	2.4	—	—	32.6	—	—	—	29.2	—	—
	4.8	—	—	40.2	—	—	—	45.9	—	—
−30°	1.2	45.1	140.0	19.1	185.0	34.7	139.1	13.8	—	—
	2.4	59.1	—	36.3	—	51.6	—	36.0	—	—
	4.8	72.1	—	53.7	—	65.5	—	55.9	—	—
−35°	1.2	41.5	178.3	6.0	125.0	32.5	169.5	8.6	155.4	—
	2.4	57.6	—	23.8	—	53.4	—	21.2	—	—
	4.8	67.1	—	47.1	—	71.3	—	34.9	—	—
−40°	1.2	37.5	232.8	12.3	165.0	32.1	341.0	8.5	209.6	8.3
	2.4	53.0	—	20.7	—	39.6	—	16.7	—	21.1
	4.8	69.3	—	45.2	—	50.1	—	43.0	—	35.2
−45°	1.2	29.4	244.7	9.7	195.0	20.0	284.0	11.1	248.7	—
	2.4	50.9	—	18.8	—	40.1	—	16.0	—	—
	4.8	61.9	—	52.8	—	62.2	—	52.1	—	—
−50°	1.2	30.7	323.0	9.7	260.0	24.3	301.0	12.3	299.0	1.7
	2.4	56.0	—	18.6	—	39.6	—	16.3	—	3.2
	4.8	55.8	—	55.5	—	61.1	—	51.5	—	7.6
−55°	1.2	—	—	25.2	340.0	—	—	20.1	375.0	0
	2.4	42.8	—	43.5	—	39.5	—	40.3	—	0
	4.8	60.6	—	63.5	—	54.1	—	60.5	—	0
−60°	1.2	34.9	705.0	11.4	300.0	26.1	593.0	11.5	297.0	0
	2.4	47.2	—	26.7	—	42.0	—	18.1	—	0
	4.8	56.8	—	42.8	—	56.1	—	34.7	—	0
−65°	1.2	33.8	41.0	15.5	310.0	42.0	718.0	12.2	295.0	—
	2.4	50.0	—	20.6	—	54.2	—	18.9	—	—
	4.8	70.1	—	41.5	—	—	—	38.7	—	—

[Isobutylene] = 3.0 M, [Et$_2$AlBr] = 6.0 × 10^{-3} M, MeX = 15 ml, 20–60 min.

system in the range from −25 °C to −65 °C. Only the yield data will be discussed in this section.

According to the data, *t*-BuCl and *t*-BuBr were equally reactive initiators. Comparable yields were obtained using MeCl and MeBr solvents, particularly at high *t*-BuX concentration (4.8×10^{-4} M). At lower *t*-BuX concentration (1.2×10^{-4} and 2.4×10^{-4} M), yields were generally lower with *t*-BuBr than with *t*-BuCl, and for a given initiator with MeBr than MeCl. Interestingly, *t*-BuI in conjunction with Et$_2$AlBr initiated polymerization at −40 °C and −50 °C. However, below −50 °C, polymerization did not take place. Polymerization did not occur also using MeI solvent.

The effect of temperature on yields was insignificant in the range from −25 °C to −55 °C. At and below −60 °C, yields dropped slightly while polymerization did not occur at −75 °C. The floor temperature for these systems is about −65 °C.

Initiator efficiencies calculated for [*t*-BuX] = 2.4×10^{-4} M at −60 °C (Table 8) decreased as: *t*-BuCl/Et$_2$AlBr/MeCl > *t*-BuBr/Et$_2$AlBr/MeCl > > *t*-BuCl/Et$_2$AlBr/MeBr > *t*-BuBr/Et$_2$AlBr/MeBr > *t*-BuI/Et$_2$AlBr/MeCl = 0. Thus, initiator reactivity decreased as *t*-BuCl > *t*-BuBr > *t*-BuI and for a given initiator, reactivity was solvent dependent as MeCl > MeBr > MeI = 0.

4. Effect of *t*-BuX, MeX and Temperature on Isobutylene Polymerization Using Et$_2$AlI Coinitiator

Since neither *t*-BuI[34] nor I$_2$[15] are initiators in conjunction with R$_3$Al or Et$_2$AlCl, the use of Et$_2$AlI becomes important from the point of view of introducing -I in the counteranion. Recently Italian workers[23] also used this approach and studied isobutylene polymerization with I$_2$/Et$_2$AlI initiator system.

Table 6 gives the data obtained with *t*-BuX/Et$_2$AlI initiator system using MeCl solvent. Et$_2$AlI was found to induce polymerization even in the absence of purposefully added *t*-BuX. As these experiments were carried out in a dry box, traces of moisture (< 30 ppm) may have functioned as protogenic initiators. The data obtained in these experiments, *i.e.* in the absence of *t*-BuX, are shown in columns 9 and 10 in Table 6.

In conjunction with Et$_2$AlI, for the first time it was possible to compare the reactivities of *t*-BuCl, *t*-BuBr and *t*-BuI initiators. According to the data in Table 6 these initiators exhibit similar efficiencies in the range from −30° to −65 °C. However, the rate of polymerization was lower with *t*-BuCl than with *t*-BuBr or *t*-BuI.

The rate of polymerization was higher in the presence than in the absence of *t*-BuX. Decreasing temperatures strongly reduced conversion and rates in the absence of *t*-BuX, particularly at or below −50 °C. In contrast, conversions and rates remained fairly constant when *t*-BuX was added in the range from −30 °C to −65 °C. The floor temperature was ∼ −75 °C for *t*-BuX/Et$_2$AlI/MeCl system.

Similarly to Et$_2$AlCl and Et$_2$AlBr systems, polymerization did not occur with Et$_2$AlI using MeI. Surprisingly, however, MeBr was also a poison and polymer did not form in MeBr. These observations are further discussed in Section VII.

Table 6. Effect of temperature on yield and $\bar{M}_v$ of PIB prepared with t-BuX/Et$_2$AlI and "H$_2$O"/Et$_2$AlI systems using MeCl

Temp. °C	t-BuX $\times 10^4$, M	t-BuCl Yield %	t-BuCl $\bar{M}_v \times 10^{-3}$	t-BuBr Yield %	t-BuBr $\bar{M}_v \times 10^{-3}$	t-BuI Yield %	t-BuI $\bar{M}_v \times 10^{-3}$	"H$_2$O" Yield %	"H$_2$O" $\bar{M}_v \times 10^{-3}$
−30°	2.4	85.4	258.0	85.5	130.7	–	–	86.2	182.0
−40°	1.2	–	–	–	–	84.3	103.0	–	–
	2.4	86.5	172.0	83.4	194.0	88.8	118.3	75.2	293.0
−45°	1.2	–	–	–	–	87.8	126.0	–	–
	2.4	89.9	–	89.3	–	88.4	164.0	86.5	–
−50°	1.2	–	–	–	–	80.9	277.0	–	–
	2.4	88.6	280.0	88.8	286.0	85.9	160.0	57.7	411.0
−55°	1.2	–	–	–	–	87.8	273.5	–	–
	2.4	86.6	335.2	87.9	408.5	86.5	261.5	39.2	507.0
−60°	1.2	–	–	–	–	36.4	280.0	–	–
	2.4	67.8	488.5	85.3	509.5	88.9	393.0	8.9	172.0
−65°	1.2	–	–	–	–	65.0	326.7	–	–
	2.4	71.0	471.0	78.9	359.0	87.4	447.3	3.6	80.0
−70°	1.2	–	–	–	–	4.2	–	–	–
	2.4	9.6	231.0	55.0	378.0	8.9	–	0	–
−75°	1.2	–	–	–	–	1.3	–	–	–
	2.4	1.0	–	3.7	43.1	2.3	–	0	–

[Isobutylene] = 3.0 M, [Et$_2$AlI] = 6.0 $\times 10^{-3}$ M, ["H$_2$O"] = unknown, trace quantity, MeCl = 15 ml, Time = 20–60 mins.

Initiator efficiencies (Table 8) calculated for yields obtained at $-60\,°C$ decreased in the order t-BuI $>$ t-BuBr $>$ t-BuCl. Initiator reactivity based on initiator efficiency, rates of polymerization and floor temperature decreased as: t-BuI $>$ t-BuBr $>$ t-BuCl and depending on solvents as MeCl $\gg$ MeBr, MeI = 0.

5. Isobutylene Polymerization Using "H_2O"/$EtAlCl_2$ Initiator and n-Pentane Solvent

$EtAlCl_2$ coinitiator induces isobutylene polymerization without purposeful addition of an initiator. In this respect, $EtAlCl_2$ is similar to conventional Lewis acids like $AlCl_3$, BF_3 or $AlBr_3$. Kennedy and Squires[35] compared the molecular weights of PIB prepared with $EtAlCl_2$, BF_3 and $AlCl_3$ using MeCl solvent. Marek and co-workers[16] polymerized isobutylene with $EtAlCl_2$ using n-heptane solvent. They found that the rates were very high and $\Delta E_{Rate} = 0$ in the range from $21°$ to $-55\,°C$. The initial rate of polymerization was first order in monomer and second order in $EtAlCl_2$.

Table 7 gives conversions and molecular weights obtained at various temperatures. Initiation was instantaneous even in the absence of t-BuX. The polymerization was homogeneous in n-pentane over the whole temperature range. The effect of temperature on conversion was insignificant in the range from $-30°$ to $-102\,°C$. The high reactivity of $EtAlCl_2$ is indicated by the high rates and nearly complete conversions obtained using even a nonpolar medium and low temperatures.

Table 7. Effect of temperature on yield and $\overline{M}_v$ of PIB prepared with "H_2O"/$EtAlCl_2$/n-pentane system

Temp. °C	Yield %	$\overline{M}_v$ $\times 10^{-3}$
-30	81.9	40.5
-40	82.8	92.5
-50	85.5	138.8
-60	86.0	276.0
-70	85.6	452.5
-80	92.0	2,125.0
-90	79.5	2,178.0
-102	69.7	2,654.0

[Isobutylene] = 3.0 M.
[$EtAlCl_2$] = 6.1×10^{-3} M.
Pentane = 15 ml, 20 mins.

VII. Effect of MeI and MeBr on Isobutylene Polymerization

Kennedy et al.[1] found that the rate of the t-BuX + Me_3Al model reaction is strongly affected by the nature of the halogen in MeX solvent, specifically the rates decreased as MeCl > MeBr > MeI. Subsequently, Kennedy and Rengachary[11] studied isobutylene polymerizations and found that conversions and molecular weights were lower using MeBr than MeCl. According to the data presented in Section VI isobutylene could not be polymerized with t-BuX/Et_2AlX/MeI or t-BuX/Et_2AlI/MeBr systems. Also, with t-BuX/Me_3Al and t-BuX/Et_2AlCl, PIB yields were distinctly lower using MeBr than MeCl. In view of the mechanistic implication of this poisoning effect of MeI and MeBr on isobutylene polymerization, it was decided to investigate these findings in detail.

MeI poisoning in isobutylene polymerization with t-BuX/Me_3Al systems was studied using mixed solvent systems containing various proportions of MeCl and MeI. Results, together with experimental details and materials used are reported in Fig. 2.

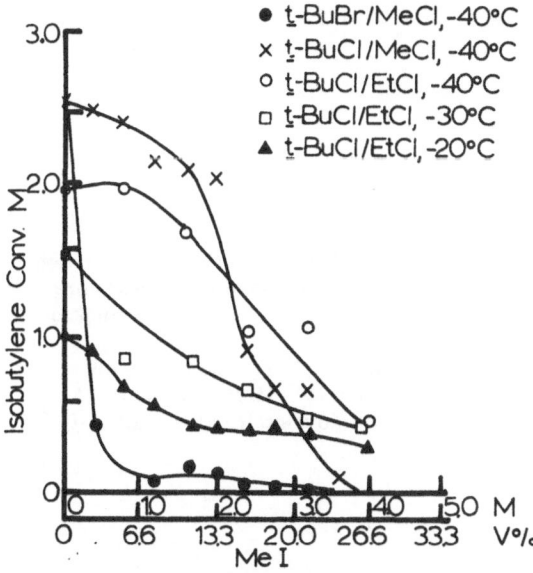

Fig. 2. Effect of MeI concentration on conversion of isobutylene polymerized using t-BuX/Me_3Al initiator system
([Isobutylene] = 3.0 M, [Me_3Al] = 2.5 x 10^{-2} M, [t-BuX] = 8.8 x 10^{-3}M in MeCl, [t-BuX] = = 4.5 x 10^{-3} M in EtCl, MeI + MeCl or EtCl = 15 ml, 60 mins.)

The increase of MeI concentration in mixed solvent strongly decreased conversion and the magnitude of this effect was dependent on t-BuX initiator and temperature used. Thus, poisoning was more pronounced with t-BuBr than with t-BuCl and it increased with decreasing temperatures. In experiments with EtCl/MeI mixtures, [MeI] = 4 M (26.6 vol.%) reduced the yield by about 70% at −20° and −30 °C, 80% at −40 °C, and 100% at −50 °C. The increase in conversion with decrease in temperature from −20° to −40 °C may be due to a decrease in the rate of termination. Similarly, using MeCl/MeI mixtures at −50 °C, [MeI] = 1.5 M (10 vol.%) completely prevented polymerization, though high conversion was obtained at −40 °C.

MeI was also a strong poison with Et_2AlCl or Et_2AlI. Thus, 0.8 M (5 vol.%) and 4 M (26.6 vol.%) MeI prevented polymerization using t-BuCl/Et_2AlCl [4.4 x 10^{-4}

and 6.0×10^{-3} M, respectively] and t-BuCl/Et$_2$AlI [1.9×10^{-4} and 6.0×10^{-3} M, respectively] initiator systems, respectively, at $-40\,°C$. [MeI concentrations were not optimized].

A similar poisoning effect with MeBr was discovered using Et$_2$AlI coinitiator. Poisoning by MeBr was studied using MeCl/MeBr mixtures with t-BuBr and "H$_2$O" initiators at different temperatures. Results are shown in Figs. 3 and 4. As with MeI,

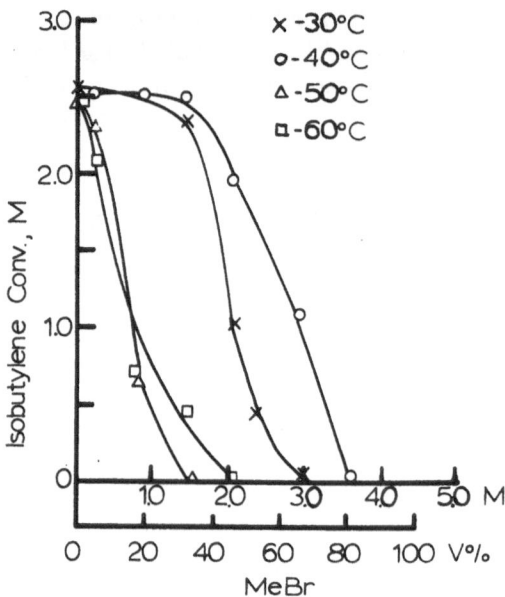

Fig. 3. Effect of MeBr concentration on conversion of isobutylene using t-BuBr/Et$_2$AlI at different temperatures
[Isobutylene] = 3.0 M, [Et$_2$AlI] = 6.0×10^{-3}M, MeCl + MeBr = 15 ml, 20–60 mins. [t-BuBr] = 2.4×10^{-4}M

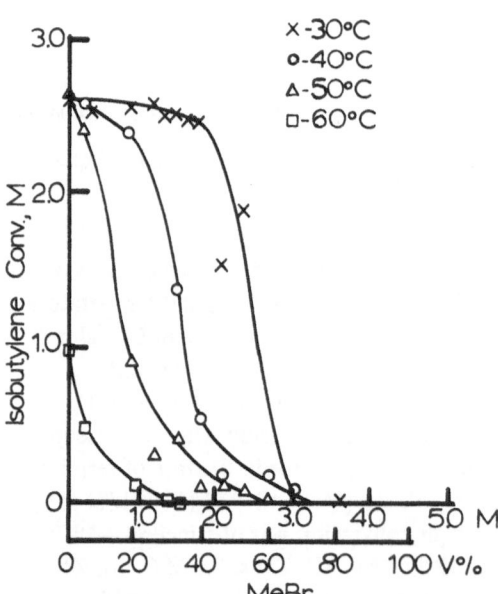

Fig. 4. Effect of MeBr concentration on isobutylene conversion using " "H$_2$O"/Et$_2$AlI system at different temperatures
[Isobutylene] = 3.0 M, [Et$_2$AlI] = 60×10^{-3}M, MeCl + MeBr = 15 ml, 20–60 mins

poisoning with MeBr increased with the less active initiator and at lower temperatures. Thus, strong poisoning was observed with "H_2O" and at $-50°$ and $-60\,°C$. For t-BuBr/Et_2AlI system, conversion was higher at $-40°$ than at $-30\,°C$, which might be due to decrease in termination. Poisoning by MeBr was very strong with the t-BuBr/Et_2AlI initiator system at $-50°$ and $-60\,°C$, $i.e.$ while conversion was unaffected at $-30°$ and $-40\,°C$, it was zero at $-50°$ and $-60\,°C$ in the presence of 2 M (45 vol.%) MeBr (Fig. 3). The effect of temperature was also dramatic for "H_2O"/Et_2AlI and poisoning increased significantly with a decrease in temperature from $-30°$ to $-60\,°C$. Evidently initiation was more affected than termination by temperature with this initiator system.

This study shows that MeI, and, for some systems, MeBr are potent poisons in isobutylene polymerization. The extent of poisoning increases in the presence of less reactive initiator and by decreasing temperatures. The mechanism of poisoning is discussed in Section VIII.

VIII. Effect of the Nature of Halogen in t-BuX Initiators, Et_2AlX Coinitiators and MeX Solvents on Isobutylene Polymerizations

The effect of t-BuX, Et_2AlX and MeX on PIB yield and polymerization rate was studied (Sections V, VI, VII). Relative initiator reactivities were determined based on yields, initiator efficiencies at $-60\,°C$, polymerization rates and floor temperatures. Initiator reactivity orders can be summarized as follows:

1) With Me_3Al and Et_2AlCl, using MeCl or MeBr, t-BuX reactivity decreases as: t-BuCl $>$ t-BuBr $\gg$ t-BuI $= 0$, and depending on the nature of solvent as MeCl $>$ MeBr $\gg$ MeI $= 0$.

2) With Et_2AlBr, using MeCl or MeBr, t-BuX reactivity decreases as t-BuCl $\geqslant$ t-BuBr $>$ t-BuI and depending on solvent as MeCl $\geqslant$ MeBr $\gg$ MeI $= 0$.

3) With Et_2AlI using MeCl, t-BuX reactivity decreases as t-BuI $\geqslant$ t-BuBr $>$ t-BuCl, and depending on nature of solvent as MeCl $\gg$ MeBr, MeI $= 0$.

Table 8. Initiator efficiencies[a] in isobutylene polymerization

Coinitiator	t-BuX M	t-BuCl MeCl	MeBr	t-BuBr MeCl	MeBr	t-BuI MeCl
Me_3Al	8.9×10^{-3}	280	160	280	0	0
Et_2AlCl	1.9×10^{-4}	9200	1200	2700	0	0
Et_2AlBr	2.4×10^{-4}	5900	3300	5300	2300	0
Et_2AlI	2.4×10^{-4}	8500	0	10,600	0	11,100

[a]   Initiator Efficiency = Monomer Conv., M/Initiator Conc., M.
      [Isobutylene] = 3.0 M, [Me_3Al] = 2.5×10^{-2} M, $-40\,°C$.
      [Et_2AlX] = 6.0×10^{-3} M, $-60\,°C$. MeX = 15 ml.

4) A comparison of initiator effiencies at $-60\,^\circ$C (Table 8) leads to the following overall efficiency sequence (initiator/coinitiator/solvent):

t-BuI/Et$_2$AlI/MeCl > t-BuBr/Et$_2$AlI/MeCl > t-BuCl/Et$_2$AlCl/MeCl > t-BuCl/ Et$_2$AlI/MeCl > t-BuCl/Et$_2$AlBr/MeCl > t-BuCl/Et$_2$AlBr/MeBr > t-BuBr/Et$_2$AlBr/ MeCl > t-BuBr/Et$_2$AlCl/MeCl > t-BuBr/Et$_2$AlBr/MeBr > t-BuCl/Et$_2$AlCl/MeBr $\gg$ $\gg t$-BuBr/Et$_2$AlCl/MeBr, t-BuX/Et$_2$AlI/MeBr, t-BuI/Et$_2$AlCl/MeX, t-BuI/Et$_2$AlBr/MeX, t-BuX/Et$_2$AlX/MeI = 0. This sequence is somewhat different at higher temperatures.

5) Based on overall initiator efficiencies, coinitiator efficiencies seem to decrease as: Et$_2$AlI > Et$_2$AlBr > Et$_2$AlCl. The coinitiator efficiency sequence for the alkylaluminum families is EtAlCl$_2$ > Et$_2$AlX > Me$_3$Al.

6) Highest PIB yields were obtained using MeCl. Polymerization could not be carried out using MeI and with Et$_2$AlI using MeBr.

This very large amount of information can be readily rationalized on the basis of a mechanism developed to explain the results of cationic model initiation and termination experiments[10, 34]. It is postulated that the initiator efficiencies in isobutylene polymerization using t-BuX/Et$_2$AlX/MeX systems are mainly determined by the overall rate of initiation, R_{int}. Initiation according to numerous studies[10, 11, 34] involves the following steps:

Complexation: $[Et_2AlX]_2 + 2\,MeX \overset{K}{\rightleftharpoons} 2\,Et_2AlX \leftarrow MeX$

Displacement: $Et_2AlX \leftarrow MeX + t\text{-}BuX \overset{k_1}{\rightleftharpoons} Et_2AlX \leftarrow t\text{-}BuX + MeX$

Ionization: $Et_2AlX \leftarrow t\text{-}BuX \overset{k_2}{\rightleftharpoons} Et_2AlX_2^{\ominus}\,t\text{-}Bu^{\oplus}$

Initiation: $t\text{-}Bu^{\oplus}\,Et_2AlX_2^{\ominus} + i\text{-}C_4H_8 \rightarrow t\text{-}Bu\text{-}i\text{-}C_4H_8^{\oplus}\,Et_2AlX_2^{\ominus} \overset{i\text{-}C_4H_8}{\longrightarrow} Propagation$

where K is the equilibrium constant for complexation and k_1 and k_2 are the rate constants for displacement and ionization, respectively.

In view of the chemical nature of alkylaluminums and methyl halides, complexation is most likely to be rapid and complete, $i.e.$ K is large. Indeed Me$_3$Al and a variety of Lewis bases were found to complex rapidly[2]. Initiation, $i.e.$, t-butyl cation attack on monomer, is also rapid since it is an ion molecule reaction which requires very little activation energy. Thus, it appears that R_{int} and hence initiator reactivity are determined by the rate of displacement R_1 and ionization R_2.

The rate of the $t\text{-}BuX + Me_3Al \overset{MeX}{\longrightarrow} t\text{-}BuMe + Me_2AlX$ reaction decreases as X = Cl > Br > I[10]. This decrease is explained by a decrease in the rate of displacement of MeX by t-BuX, which in turn is determined by the basicity and/or size of the halogen in t-BuX. Since the basicity decreases and size increases as X changes from Cl to Br to I, the rate of displacement, R_1, decreases. In isobutylene polymerization using t-BuX/Me$_3$Al/MeX and t-BuX/Et$_2$AlCl/MeX (X = Cl, Br, I), the t-BuX reactivity decreases as t-BuCl > t-BuBr $\gg$ t-BuI = 0. The similarity between initiator reactivity sequences in model and polymerization reactions indicates that the rate governing event is the same for both, $i.e.$, the rate of displacement, R_1.

The rate of ionization, R_2, is determined by the thermodynamic stability of the counteranion $Et_2AlX_2^{\ominus}$, which in turn is governed by charge delocalization on the counterion. For Et$_2$AlI systems, charge delocalization and hence counterion stability decreases as $Et_2AlI_2^{\ominus} > Et_2AlIBr^{\ominus} > Et_2AlICl^{\ominus}$, which follows the sequence of initiator efficiencies found; $i.e.$, t-BuI > t-BuBr > t-BuCl. Also, considering the

Lewis acidities of Et_2/AlX), the displacement of MeX by t-BuX is expected to be slowest (rate determining) with the strongest Lewis acid Et_2AlCl and fastest (non-controlling) with the weaker Lewis acid Et_2AlI. Schematically,

Rate Determining Event

Displacement		Ionization
R_1, Et_2AlCl	$<$	R_2, Et_2AlCl
$\wedge$		$\wedge$
R_1, Et_2AlI	$>$	R_2, Et_2AlI

For Et_2AlBr-based systems, rates of displacement and ionization are comparable so that polymerization rates and initiator reactivities obtained with t-BuCl and t-BuBr are similar. For t-BuI/Et_2AlBr, R_1 may be rate determining since initiation takes place only at relatively high temperatures ($-40\ °C$).

The coinitiator reactivity sequence of alkylaluminum compounds, $EtAlCl_2 > Et_2AlX > Me_3Al$, may similarly be due to differences in the rates R_1 and R_2.

The nature of halogen in the MeX solvent also profoundly affects initiator efficiency and decreases as: $MeCl > MeBr \gg MeI = 0$. Polymerization did not occur in MeI and also in MeBr when Et_2AlI-based initiator systems were used. At least two hypotheses may account for this sequence.

i) The strength of the coordinate bond $Et_2AlX \leftarrow MeX$ increases from MeCl to MeBr to MeI, $i.e.$, as the polarizability of halogen in MeX increases, strong coordination of MeX reduces the rate of displacement, R_1, and leads to decreased initiator efficiency.

ii) Poisoning is due to the formation of propagation-inactive alkylhalonium ions. Alkyl halides are known to form stable halonium ions with incipient carbenium ions[38].

$$G^\ominus \equiv (CH_3)_3AlX^\ominus \text{ (or } Et_2AlX_2^\ominus)$$

Thus, in the systems under consideration, MeX may form halonium ions with growing carbenium ions. Since the stability of halonium ions depends on the polarizability of the halogen[38], $-I > -Br > -Cl$, MeI should form the most stable halonium ions, *i.e.*, have most pronounced poisoning effect, followed by MeBr and MeCl. Indeed, MeI may even compete for the carbocation with highly nucleophilic counterions.

Like carbenium ions, halonium ions may undergo propagation, transfer or termination. The significant decrease in monomer conversion in the presence of MeI and MeBr indicates that termination becomes important. Halonium ion formation also explains more pronounced poisoning with less efficient initiators, *i.e.*, *t*-BuBr or "H_2O" and at lower temperatures, *i.e.*, $-50°$ or below. It seems halonium ion formation is greatly favored by the decreased concentration of incipient carbenium ions, under these conditions.

MeBr is a strong poison only with Et_2AlI coinitiator. Since Et_2AlI forms the least nucleophilic counterion, $Et_2AlIX^{\ominus}$, it is expected to produce a relatively free carbenium ion, facilitating bromonium ion formation by interaction with MeBr solvent. With more nucleophilic counteranions, like $Me_3AlX^{\ominus}$ or $Et_2AlX_2^{\ominus}$ (X = Cl, Br), bromonium ion formation is more difficult and poisoning is modest. Evidently, the less stable bromonium ions form only with weakly nucleophilic counterions. MeCl is the weakest poison or may be inert, since chloronium ions are highly unstable.

Lastly, the effect of temperature on PIB yield for the t-BuX/Et_2AlX/MeX systems can be explained. In general, the yield is unaffected or decreases somewhat with decrease in temperature. At a certain temperature, yield drops to zero. Such an effect of temperature indicates, in agreement with our assumption, that the yield is dependent on the rate of initiation.

The possibility of halide scrambling between Et_2AlX and t-BuX and MeX exists and should be considered. Halide exchange between Et_2AlX and t-BuX is probably negligible under our experimental conditions *i.e.*, at very low t-BuX concentrations and considering that t-BuX was the last ingredient added to the system containing a large excess of isobutylene. Halide exchange between Et_2AlX and MeX has been described (39, 40) to occur readily at 80 °C between MeCl and Et_2AlBr or Et_2AlI. Similarly, complete halogen exchange occurs in MeCl-$AlBr_3$, MeCl—AlI_3 and MeBr-AlI_3, systems at $0°$ in 24 hrs, however, the reverse reaction, *i.e.*, exchange between MeI-$AlCl_3$, MeBr-$AlCl_3$ and MeI-$AlBr_3$, does not take place[41]. Evidently, halide exchange occurs preferentially when the MeX contains the more electronegative (or the Al compound contains the less electronegative) halide.

In view of these facts the following exchanges may occur in the systems under investigation:

$$MeCl + Et_2AlBr \rightleftharpoons MeBr + Et_2AlCl$$
$$MeCl + Et_2AlI \rightleftharpoons MeI + Et_2AlCl$$
$$MeBr + Et_2AlI \rightleftharpoons MeI + Et_2AlBr$$

A detailed examination of these systems leads us to conclude that these halide exchanges are most likely absent or negligible under our conditions, *i.e.*, low temperatures (from $-30°$ to $-75 °C$) and < 1 hr. The *in situ* formation of Et_2AlCl (cf. the first two equations) from Et_2AlBr and Et_2AlI in the presence of t-BuCl would yield results similar to those obtained with the t-BuCl/Et_2AlCl system. In fact, the differences obtained in initiator reactivity, $\Delta E_{\bar{M}_v}$ values, molecular weights and molecular weight distributions for the t-BuCl/Et_2AlCl and t-BuCl/Et_2AlBr or t-BuCl/Et_2AlI systems, all using MeCl solvent, are very large indicating that these initiators are quite dissimilar. In the same vein, t-BuI and "H_2O" do not function as initiators in conjunction with Et_2AlCl but are initiators in the presence of Et_2AlI suggesting that Et_2AlCl is not formed from MeCl + Et_2AlI. That Et_2AlBr is not formed from MeBr + Et_2AlI is shown by the fact that while MeBr is a satisfactory solvent for Et_2AlBr-based systems it acts as a poison for Et_2AlI-based systems. These results and considerations suggest that halide exchange is most likely absent in the systems under study.

In sum, the effect of t-BuX initiators, Et_2AlX coinitiators, MeX solvents and temperature on PIB yield has been investigated. As the halogen changes from $-Cl$ to $-Br$ to $-I$ in t-BuX or MeX, initiator reactivities decrease, except for the Et_2AlI system. Thus, initiator reactivities decrease as t-BuCl $> t$-BuBr $> t$-BuI and depending on the solvent as MeCl $>$ MeBr $\gg$ MeI $= 0$. For Et_2AlI systems, t-BuX reactivities decrease as t-BuI $> t$-BuBr $> t$-BuCl. Polymerization could not be carried out using MeI, or using MeBr with Et_2AlI coinitiator. Initiator reactivities also decreased by changing the halogen from $-I$ to $-Br$ to $-Cl$ in Et_2AlX coinitiators. For the alkyl-aluminum families, the initiator reactivities decrease as $EtAlCl_2 > Et_2AlX > Me_3Al$.

Initiator reactivity orders can be explained on the basis of differences in the rate of displacement of MeX from $Et_2AlX \leftarrow$ MeX complexes by t-BuX and/or the rate of ionization of $Et_2AlX \leftarrow t$-BuX complexes. PIB yields decrease with increase of MeI or MeBr concentration. This poisoning effect has been attributed to the formation of propagation-inactive halonium ions.

The effect of temperature on PIB yield shows the existence of a floor temperature, *i.e.*, a temperature below which initiation does not take place.

On the basis of initiator efficiencies calculated from yields obtained at $-60 °C$, an overall initiator efficiency order has been developed for t-BuX/Et_2AlX/MeX systems.

The molecular weight data in this paper will be discussed in detail in the subsequent paper[1].

IX. Conclusions

Isobutylene polymerization has been carried out using t-BuX initiators, Me_3Al, Et_2AlX and $EtAlCl_2$ coinitiators, MeX and n-pentane solvents at different temperatures. The effects of t-BuX, Me_3Al, Et_2AlX and $EtAlCl_2$, MeX and temperature on polymerization rate, PIB yield and molecular weight as well as molecular weight distribution have been investigated in great detail. Also, the effects of initiators, coinitiators, solvents and temperature on $\Delta E_{\bar{M}_v}$ and in some cases $\Delta E_{\bar{M}_n}$ and $\Delta E_{\bar{M}_w}$ have been evaluated.

Using the t-BuX/Me$_3$Al/MeX system, a preferred reagent addition sequence has been found to be i-C$_4$H$_8$/MeX/Me$_3$Al/t-BuX. This sequence has been used in these investigations. Based on polymerization rates at $-40\,^\circ$C, overall polymer yields, floor temperature and initiator efficiencies at $-40\,^\circ$C, overall initiator reactivity is found to decrease as t-BuCl $>$ t-BuBr $\gg$ t-BuI $= 0$ and initiator reactivity is dependent on solvent as MeCl $>$ MeBr $\gg$ MeI $= 0$. Similarity of reactivity sequences in isobutylene polymerization and in cationic model initiation and termination studies[13] suggest that initiator reactivities are determined by the rate of initiation, R_1.

A detailed investigation of isobutylene polymerizations using t-BuX/Et$_2$AlX/MeX has shown the following initiator reactivity sequences:
with Et$_2$AlCl: t-BuCl $>$ t-BuBr $\gg$ t-BuI $= 0$ and depending on solvent,
MeCl $>$ MeBr $\gg$ MeI $= 0$;
with Et$_2$AlBr: t-BuCl $\geqslant$ t-BuBr $>$ t-BuI and depending on solvent, MeCl $\geqslant$ MeBr $\gg$
$\gg$ MeI $= 0$;
with Et$_2$AlI: t-BuI $\geqslant$ t-BuBr $>$ t-BuCl and depending on solvent, MeCl $>$ MeBr $\gg$
$\gg$ MeI $= 0$.
Based on alkylaluminum families, initiator reactivity is found to decrease as
EtAlCl$_2$ $>$ Et$_2$AlX $>$ Me$_3$Al and for Et$_2$AlX systems as, Et$_2$AlI $>$ Et$_2$AlBr $>$
Et$_2$AlCl. Based on initiator efficiencies at $-60\,^\circ$C, an overall initiator/coinitiator/solvent reactivity sequence has been developed.

These reactivity sequences have been explained by considering initiation to involve four steps, *e.g.* complexation, displacement, ionization and initiation; and subsequently proposing that displacement or ionization is rate controlling. The initiator reactivity order is shown to change depending on whether displacement or ionization becomes rate governing.

The absence of polymerization using MeI and using MeBr for Et$_2$AlI system has been further investigated using mixtures of MeCl and MeI or MeBr. This had led to the proposition that poisoning by MeI or MeBr is due to the formation of propagation-inactive halonium ions.

Acknowledgements. The authors are grateful to Drs. D. McIntyre and S. Shih of The University of Akron for the fractionated PIB samples. Financial help by the National Science Foundation and the Firestone Tire and Rubber Company is gratefully acknowledged.

X. References

[1] Kennedy, J. P., Trivedi, P.D.: Adv. Polymer Sci. *28*, 31 (1978)
[2] Kennedy, J. P.: Belgian Patent *663*, 319 (April 30, 1965)
[3] Kennedy, J. P., Melby, E. M.: J. Polymer Sci. *13*, 29 (1975)
[4] Kennedy, J. P., Smith, R. R.: Polymer Preprints *13* (2), 710 (1972)
[5] Kennedy, J. P., Charles, J. J., Davidson, D. L.: Polymer Preprints *14* (2), 974 (1973)
[6] Sigwalt, P., Palton, A., Miskovic, M.: J. Polymer Sci. *C, 56*, 13 (1976)
[7] Vidal, A., Kennedy, J. P.: J. Polymer Sci. *B, 14*, 489 (1976)
[8] Kennedy, J. P.: Inter. Symp. Macromol. Chem., Tokyo, Kyoto, Abstract 2.104 (1966)
[9] Kennedy, J. P.: J. Polymer Sci. *A-1, 6*, 3139 (1968)
[10] Kennedy, J. P., Desai, N. V., Sivaram, S.: J. Am. Chem. Soc. *95*, 6386 (1973)
[11] Kennedy, J. P., Rengachary, S.: Adv. in Polymer Sci. *14*, 1 (1974)
[12] Kennedy, J. P. in: Copolymerization. Chapter V, Ham, G. E. (ed.). New York: Wiley-Interscience Publ. 1964

13) Kraus, C. A.,: U. S. Patent 2,220, 930 (1940)
14) Kennedy, J. P., Gillham, J. K.: Adv. in Polymer Sci. *10*, 1 (1972)
15) Kennedy, J. P., Sivaram, S.: J. Macromol. Sci.-Chem. *A7*, 969 (1973)
16) Maslinska-Solich, J., Chmelir, M., Marek, M.: Collection Czechoslov. Chem. Commun. *34*, 2611 (1969)
17) Priola, A., Cesca, S., Ferraris, G., Milanese, S. D., Boy, M. S., Giusti, P.: U. S. Patent 3,850, 894 (1974)
18) Priola, A., Cesca, S., Ferraris, G., Milanese, S. D.: U. S. Patent 3,850, 895 (1974)
19) Priola, A., Cesca, S., Ferraris, G., Milanese, S. D., Boy, M. B., Giusti, P.: U. S. Patent 3,850, 896 (1974)
20) Priola, A., Cesca, S., Ferraris, G., Milanese, S. D.: U. S. Patent 3,850, 897 (1974)
21) Priola, A., Ferraris, G., de Maina, M., Giusti, P.: Makromol. Chem. *176*, 2271 (1975)
22) Priola, A., Cesca, S., Ferraris, G., de Maina, M.: Makromol. Chem. *176*, 2289 (1975)
23) Giusti, P., Priola, A., Magagnini, P., Narducci, P.: Makromol. Chem. *176*, 2303 (1975)
24) Cesca, S., Giusti, P., Magagnini, P., Priola, A.: Makromol. Chem. *176*, 2319 (1975)
25) Cesca, S., Priola, A., Bruzzone, M., Ferraris, G., Giusti, P.: Makromol. Chem. *176*, 2339 (1975)
26) de Maina, M., Cesca, S., Giusti, P., Ferraris, G., Magagnini, P. L.: Makromol. Chem. *178*, 2223 (1977)
27) Magagnini, P. L., Cesca, S., Giusti, P., Priola, A., de Maina, M.: Makromol. Chem. *178*, 2235 (1977)
28) Flory, P. J.: J. Am. Chem. Soc. *65*, 372 (1943)
29) Kennedy, J. P.: Cationic polymerization of olefins: A critical inventory. New York: Wiley-Interscience Publ. 1975
30) Charadame, H., Sigwalt, P.: Compt. Rend. *259*, 4273 (1964)
31) Kennedy, J. P., Langer, A. W.: Adv. in Polymer Sci. *3*, 508 (1964)
32) Kennedy, J. P., Sivaram, S.: J. Org. Chem. *38*, 2262 (1973)
33) Kennedy, J. P.: Brit. Patent 1,094, 728 (1965)
34) Kennedy, J. P., Trivedi, P. D.: Polymer Preprints *17* (2), 791 (1976)
35) Kennedy, J. P., Squires, R. G.: Polymer *6*, 579 (1965)
36) Pepper, D. C.: Quart. Rev. *8*, 90 (1954)
37) Sawada, H.: J. Macromol. Sci.-Revs. *C7*, 161 (1972)
38) Olah, G. A.: Halonium ions. New York: Wiley-Interscience Publ. 1975
39) Dahlig, W., Pasynkiewicz, S., Meszorer, L.: Roczniki Chem. *34*, 1519 (1960)
40) Pasynkiewicz, S., Dahlig, W., Meszorer, L.: Roczniki Chem. *35*, 1301 (1961)
41) Brown, H. C., Wallace, W. J.: J. Am. Chem. Soc. *75*, 6279 (1953)

Received January 18, 1978
H.-J. Cantow (editor)

Cationic Olefin Polymerization Using Alkyl Halide Alkylaluminium Initiator Systems

II. Molecular Weight Studies

Joseph P. Kennedy and Prakash D. Trivedi*

Institute of Polymer Science, The University of Akron, Akron, Ohio 44325, U.S.A.

The effect of t-BuX, Et_2AlX and $EtAlCl_2$, MeX and temperature on PIB molecular weight has been studied by GPC and viscometry. A large amount of $\overline{M}_n$, $\overline{M}_w$, MWD as well as $\overline{M}_v$ and $\Delta E_{\overline{M}_v}$ (activation energy of viscosity average molecular weight) data were generated. Influence of monomer concentration on $\overline{M}_n$ and $\overline{M}_v$ was analyzed by Mayo plots and relative rate constants were calculated. Predominant molecular weight controlling mechanisms in t-BuCl and t-BuBr/Et_2AlCl/MeCl systems were determined.

A relation between $\Delta E_{\overline{M}_v}$ and molecular weight controlling mechanisms was discovered and the effect of initiator system, solvent and temperature on $\Delta E_{\overline{M}_v}$ was explained. The present work has led to an understanding of the effect of counteranion on PIB molecular weight. These studies provide better insight into the detailed mechanism of isobutylene polymerization, in particular into the initiation and the molecular weight controlling events.

Table of Contents

* Prakash D. Trivedi present address: Central Research Laboratories, The Firestone Tire and Rubber Company, Akron, Ohio 44317, U.S.A.

I. Introduction

In previous papers[1, 2] we described reactivity studies of cationic isobutylene polymerization using t-butyl halide initiators, alkylaluminum coinitiators and methyl
halide solvents. The effects of these reagents as well as temperature on the overall
rate of polymerization and polyisobutylene (PIB) yield were studied and reactivity
orders were established. These results were explained by a modified initiation
mechanism based on an earlier model proposed by Kennedy and co-workers[3, 4].
This paper concerns the effects of t-butyl halide, alkylaluminums and methyl halide,
as well as temperature and isobutylene concentration on PIB molecular weights.
A preliminary report of this study has been published[2].

II. Experimental

The experimental details regarding the polymerization techniques and molecular
weight determinations are given in Part 1[1].

III. Gel Permeation Chromatography of Polyisobutylene

1. Introduction

GPC, has been extensively used in studies of molecular weights and molecular weight
distributions (MWD), and polymerization mechanisms[5]. In the field of cationic
polymerization, the mechanism of styrene[6, 7], α-methylstyrene[8] and p-methoxy-
styrene[9] polymerization have been examined by GPC, however, that of isobutylene
has not yet been studied by this technique. Kennedy, Shinkawa and Williams[10] have
compared the molecular weights of PIB prepared by γ-rays with those obtained by
$AlCl_3$, BF_3 and $EtAlCl_2$, using GPC, osmometry and viscometry.

  This section concerns GPC studies carried out to determine the molecular weights
and MWD of PIB's prepared with t-BuX/Et_2AlX/MeX systems and to investigate the
molecular weight controlling mechanisms.

2. Effect of Temperature on PIB Molecular Weight

Figure 1 and Table 1 show the data obtained using PIB's prepared with the t-BuCl/
Et_2AlCl/MeCl system in the $-30°$ to $-65\ °C$ range. PIB's prepared in the range from
$-30°$ to $-45\ °C$ have monomodal MWD while those prepared below $-45\ °C$ exhibit
bimodal distributions. MWD's broaden with decreasing polymerization temperature.

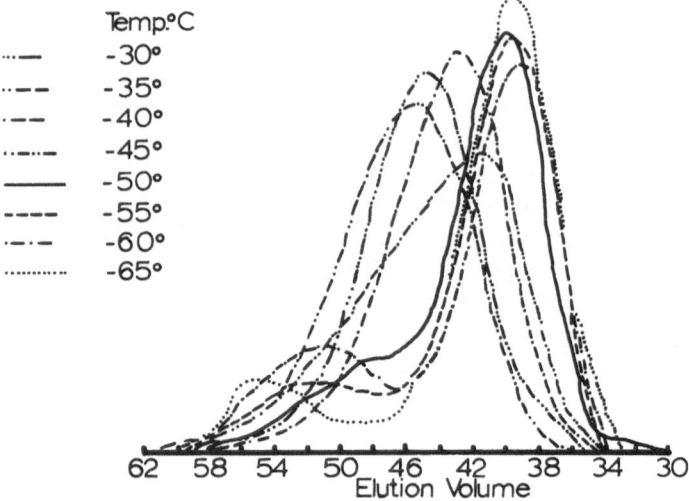

Fig. 1. GPC curves for PIB prepared with t-BuCl/Et$_2$AlCl/MeCl as a function of temperature

Table 1 gives yields and molecular weights for the total polymer and for the low and high molecular weight fractions (LMWF and HMWF). Since the resolution of GPC traces into LMWF and HMWF was estimated subjectively the molecular weights for these fractions are merely indicators of trends.

With a decrease in temperature, $\overline{M}_n$ and MWD's of the total polymer increased while a trend in $\overline{M}_w$ could not be discerned. The $\overline{M}_n$ and $\overline{M}_w$ of HMWF increased with decreasing temperatures from $-50°$ to -60 °C and decreased slightly at -65 °C. Interestingly, MWD $\simeq 1.5$ over the whole range. The $\overline{M}_n$ and $\overline{M}_w$ of LMWF decreased with some scatter with decreasing temperatures from $-50°$ to -65 °C, while MWD's remained 1.4 ± 0.1. $\Delta E_{\overline{M}_n}$ and $\Delta E_{\overline{M}_w}$ of HMWF = -1.8 kcal/mole from Arrhenius plots in $-50°$ to -65 °C range.

PIB's prepared with t-BuBr/Et$_2$AlCl/MeCl show similar characteristics. Figure 2 and Table 2 show the molecular weight data obtained in the range from $-30°$ to -60 °C. $\overline{M}_n$ and $\overline{M}_w$ of the total polymer show some scatter, though $\overline{M}_w$ and MWD tend to increase with decreasing temperatures. The data obtained at -45 °C, $i.e.$, the exceptionally high $\overline{M}_n$ and $\overline{M}_w$, are difficult to explain.

The GPC traces were monomodal in the $-30°$ to -45 °C range but exhibited bimodality below -45 °C. $\overline{M}_n$ and $\overline{M}_w$ of HMWF increased with temperature from $-50°$ to -55 °C but decreased again at -60 °C. MWD was 1.4 ± 0.1. For LMWF, $\overline{M}_n$ decreased slightly and $\overline{M}_w$ remained nearly constant with decreasing temperatures. MWD was between ~ 1.3 and ~ 1.6.

It is of interest that for the t-BuCl and t-BuBr/Et$_2$AlCl/MeCl systems most of the material is in the HMWF, $i.e.$, 80 ± 6 and $70 \pm 10\%$, respectively, and that temperature did not affect the relative amounts of HMWF and LMWF.

Figure 3 and Table 3 give the data for PIB prepared with t-BuCl/Et$_2$AlBr/MeBr in the range from $-30°$ to -65 °C. All GPC traces are monomodal, even for samples prepared below -50 °C (an exception was the sample prepared at -65 °C, which showed a small shoulder). While the molecular weight data are scattered and a trend

Table 1. GPC analysis of molecular weights of PIB prepared with t-BuCl/Et$_2$AlCl/MeCl at various temperatures

Temp. °C	Yield %	Total Polymer $\bar{M}_n$ ×10⁻³	$\bar{M}_w$	MWD	L M W F $\bar{M}_n$ ×10⁻³	$\bar{M}_w$	MWD	H M W F $\bar{M}_n$ ×10⁻³	$\bar{M}_w$	MWD	HMWF %
−30	19.2	209	406	1.9_5	—	—	—	—	—	—	—
−35	22.0	158	360	2.2_8	—	—	—	—	—	—	—
−40	20.6	302	600	1.9_8	—	—	—	—	—	—	—
−45	17.4	192	626	3.2_7	—	—	—	—	—	—	—
−50	13.8	328	1010	3.0_7	83.7	125	1.5_0	805	1190	1.4_7	82
−55	17.0	250	1090	4.3_4	52.5	78.0	1.4_8	855	1280	1.5_0	84
−60	5.4	189	1030	5.4_3	57.6	91.0	1.5_8	936	1350	1.4_5	74
−65	8.5	211	1080	5.1_2	35.8	48.1	1.3_5	825	1240	1.5_0	86

[Isobutylene] = 3.0 M, [Et$_2$AlCl] = 6.0 × 10⁻³ M, [t-BuCl] = 4.5 × 10⁻⁵ M.
MeCl = 15 ml, 20–60 Min.

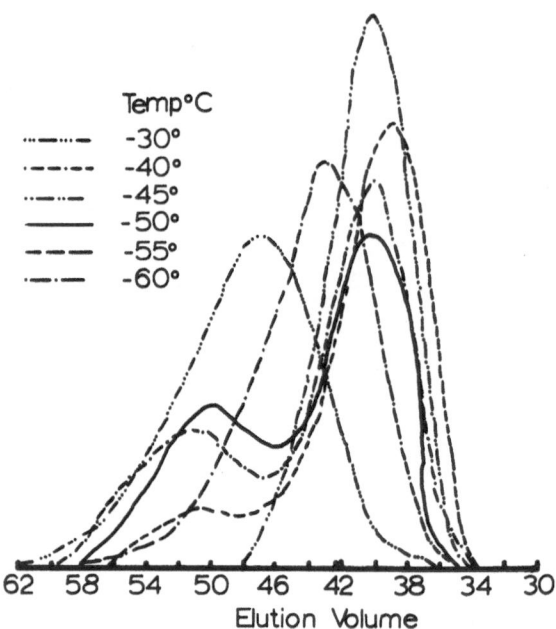

Fig. 2. GPC curves for PIB prepared with t-BuBr/Et$_2$AlCl/MeCl as a function of temperature

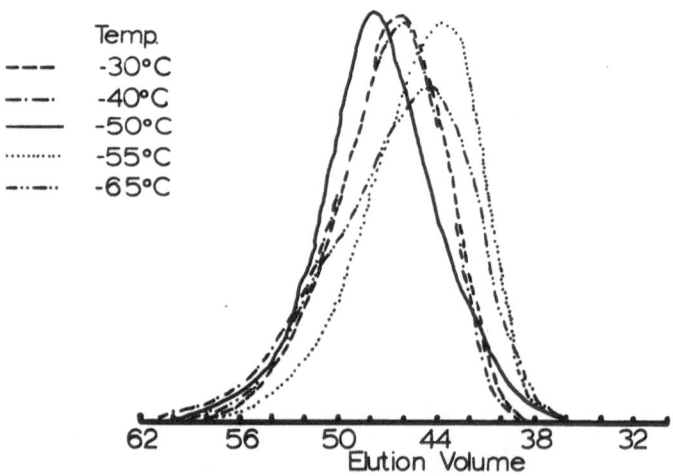

Fig. 3. GPC curves of PIB prepared with t-BuCl/Et$_2$AlBr/MeBr as a function of temperature

cannot be discerned, the MWD increases slightly from ~1.7 to ~2.3 with decreasing temperatures from −30° to −65 °C.

Data for PIB prepared using t-BuI/Et$_2$AlI/MeCl in the range from −40° to −70 °C are given in Table 3 and Fig. 4. $\overline{M}_n$ and $\overline{M}_w$ increased with decreasing temperatures and MWD $\simeq$ 2.0 to 3.0. The GPC traces were monomodal and the $\Delta E_{\overline{M}_n}$ and $\Delta E_{\overline{M}_w}$ = −4.6 kcal/mole, as calculated from Arrhenius plots of log $\overline{M}_n$ or log $\overline{M}_w$ vs. 1/T.

Table 2. GPC analysis of molecular weights of PIB prepared using t-BuBr/Et$_2$AlCl/MeCl at different temperatures

Temp. °C	Yield %	Total Polymer			L M W F			H M W F			HMWF %
		$\overline{M}_n$	$\overline{M}_w$	MWD	$\overline{M}_n$	$\overline{M}_w$	MWD	$\overline{M}_n$	$\overline{M}_w$	MWD	
		$\times 10^{-3}$			$\times 10^{-3}$			$\times 10^{-3}$			
-30	55	98	229	2.3$_3$	–	–	–	–	–	–	–
-40	13	288	606	2.1$_0$	–	–	–	–	–	–	–
-45	21	811	1120	1.3$_8$	–	–	–	–	–	–	–
-50	2.4	184	712	3.8$_8$	73	96	1.3$_2$	730	941	1.3$_1$	66
-55	17	384	1200	3.1$_5$	67	85	1.2$_7$	847	1330	1.5$_7$	89
-60	4	140	825	5.9$_0$	53	88	1.6$_1$	722	1090	1.5$_1$	68

[Isobutylene] = 3.0 M, [Et$_2$AlCl] = 6.0 x 10^{-3} M, [t-BuBr] = 4.5 x 10^{-5} M.
MeCl = 15 ml, 20–60 Min.

Table 3. GPC analysis of molecular weights of PIB prepared using t-BuCl/Et$_2$AlBr/MeBr and t-BuI/Et$_2$AlI/MeCl systems at various temperatures

Temp. °C	t-BuCl/Et$_2$AlBr/MeBr				t-BuI/Et$_2$AlI/MeCl			
	Yield %	$\bar{M}_n$	$\bar{M}_w$	MWD	Yield %	$\bar{M}_n$	$\bar{M}_w$	MWD
		$\times 10^{-3}$				$\times 10^{-3}$		
-30	19	160	273	1.7$_4$	–	–	–	–
-40	12	128	262	2.0$_5$	84	72	197	2.7$_3$
-45	–	140	–	–	88	110	223	2.0$_2$
-50	10	218	265	1.8$_9$	81	96	296	3.0$_8$
-55	25	218	431	1.9$_8$	88	186	363	1.9$_2$
-60	–	–	–	–	36	196	446	2.2$_6$
-65	16	164	375	2.2$_9$	65	267	586	2.2$_0$

[Isobutylene] = 3.0 M, [Et$_2$AlX] = 6.0 × 10^{-3}M.
[t-BuX] = 1.2 × 10^{-4} M, MeX = 15 ml.
20–60 Min.

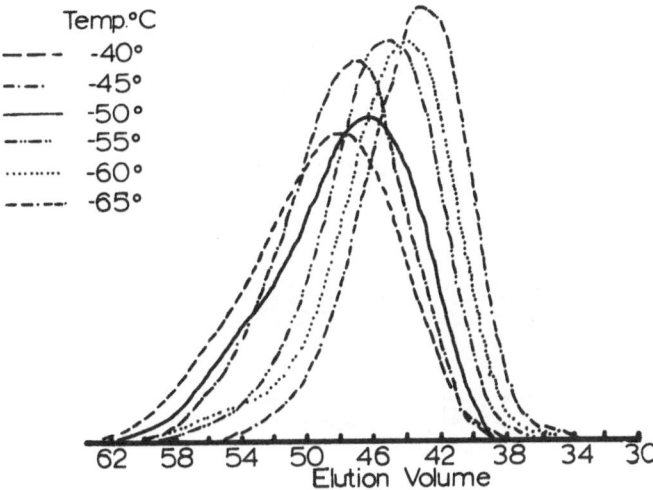

Fig. 4. GPC curves of PIB prepared with t-BuI/Et$_2$AlI/MeCl as a function of temperature

A conspicuous finding in these studies is that Et$_2$AlCl-based initiator systems lead to bimodal distributions whereas those with Et$_2$AlBr and Et$_2$AlI lead to mono-modal distributions. Also, the MWD for Et$_2$AlCl was broader ($\sim$1.4 to $\sim$5.9) than those for Et$_2$AlBr ($\sim$1.7 to $\sim$2.2) and Et$_2$AlI ($\sim$1.9 to $\sim$3.0). A possible explanation for the bimodal MWD is given in Section 5. The reason(s) for the relatively narrow MWD's ($<$2) of LMWF and HMWF remains obscure.

3. Effect of Monomer Concentration on PIB Molecular Weight

While the effect of monomer concentration [M] on $\overline{M}_v$ of PIB has been studied by a number of workers using a variety of initiator systems[11−13], such studies have not yet been carried out using $\overline{M}_n$. The objective of this phase of research was to determine the effect of [M] on $\overline{M}_n$ and thus to calculate relative rate constants of iso-butylene polymerization from GPC data.

Table 4 shows the molecular weights of PIB's prepared using the t-BuCl/Et$_2$AlCl/MeCl/n-pentane system as a function of [M] and temperature. Figure 5 shows GPC traces of PIB prepared at −60 °C.

PIB's prepared at −40 °C were monomodal with MWD = 2.2 ± 0.2 whereas those obtained at −50° and −60 °C were bimodal. For PIB's prepared at −50 °C, both $\overline{M}_n$ and $\overline{M}_w$ of the total polymer and the HMWF and LMWF increased with an increase in [M]. Increasing [M] significantly increases the relative amount of HMWF at the expense of LMWF.

The effect of [M] was similar to above at −60 °C. Thus, increasing [M] from 1.8 to 4.2 M significantly increased $\overline{M}_n$ and $\overline{M}_w$ of the total polymer and also of the two fractions. (The $\overline{M}_n$ and $\overline{M}_w$ of the total polymer and LMWF at 2.4 M were disproportionately high, probably due to the relatively low amount of LMWF.) The amount of HMWF increased from 48% to 84% with the increase in [M].

Table 4. GPC analysis of molecular weights of PIB prepared using t-BuCl/Et$_2$AlCl/MeCl systems at various monomer concentrations

$[i\text{-}C_4H_8]$ M	Yield %	$\bar{M}_v{}^a$ × 10^{-3}	Total Polymer $\bar{M}_n$ (× 10^{-3})	$\bar{M}_w$	MWD	L M W F $\bar{M}_n$ (× 10^{-3})	$\bar{M}_w$	MWD	H M W F $\bar{M}_n$ (× 10^{-3})	$\bar{M}_w$	MWD	HMWF %
-40 °C												
1.8	16.4	256	101	240	2.3_8	—	—	—	—	—	—	—
2.4	15.5	364	145	324	2.2_4	—	—	—	—	—	—	—
3.0	45.7	468	121	320	2.6_4	—	—	—	—	—	—	—
3.6	29.3	470	245	520	2.2_0	—	—	—	—	—	—	—
-50 °C												
1.8	13.8	332	80	274	3.4_0	58	107	1.8_2	419	707	1.6_9	29
3.0	15.8	563	141	525	3.7_3	74	127	1.7_2	616	862	1.4_0	54
3.6	32.0	669	198	708	3.6_2	95	186	1.9_7	734	1054	1.4_4	60
4.2	51.4	870	470	978	2.0_8	—	—	—	838	1172	1.4_0	80
-60 °C												
1.8	13.7	222	96	416	4.3_4	55	82	1.4_8	526	781	1.4_9	48
2.4	31.3	604	231	691	2.3_0	96	135	1.4_2	649	951	1.4_7	69
3.0	26.7	640	155	787	5.0_8	63	112	1.8_0	742	1146	1.5_5	65
3.6	24.9	1064	214	978	4.5_8	71	114	1.6_1	893	1266	1.4_2	73
4.2	26.3	1209	401	1148	2.6_8	96	125	1.3_0	941	1328	1.4_1	84

$[Et_2AlCl] = 6.0 \times 10^{-3}$ M, $[t\text{-}BuCl] = 4.5 \times 10^{-5}$ M.
MeCl = 13 ml, cosolvent-n-pentane.
a Data from Table 6.

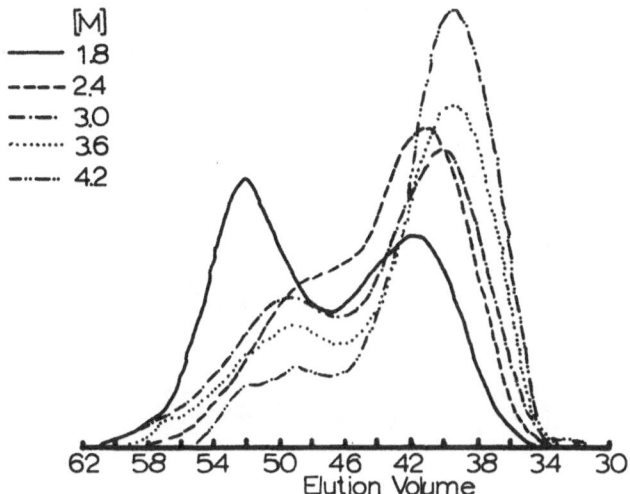

Fig. 5. GPC curves of PIB prepared with t-BuCl/Et$_2$AlCl/MeCl at $-60°$ as a function of monomer concentration

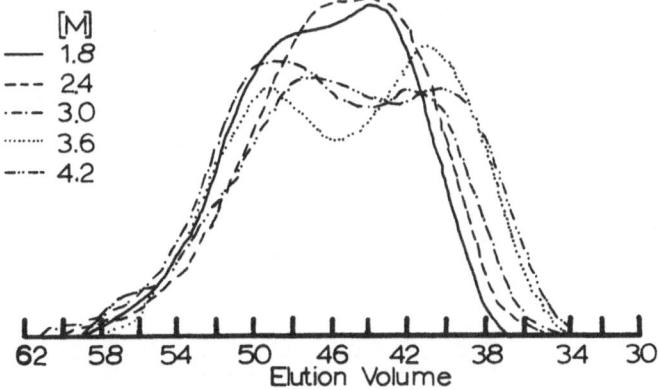

Fig. 6. GPC curves of PIB prepared with t-BuBr/Et$_2$AlCl/MeCl at $-50\,°C$ as a function of monomer concentration

According to the data shown in Table 5, the effect of [M] is similar using the t-BuBr/Et$_2$AlCl/MeCl/n-pentane system. Thus, PIB's prepared at $-30°$ and $-40\,°C$ were monomodal, whereas those obtained at $-50\,°C$ exhibited bimodality (Fig. 6). In spite of some scatter, $\bar{M}_n$ and $\bar{M}_w$ tended to increase with increasing [M] at $-30°$ and $-40\,°C$. The MWD was ~2.2 to ~2.6 and ~2.0 to ~2.8 for the samples prepared at $-30°$ and $-40\,°C$, respectively. Increasing [M] at $-50\,°C$, increased the $\bar{M}_n$ and $\bar{M}_w$ of HMWF, though the M_n and M_w of the total polymer and LMWF were unaffected. Interestingly, an increase in [M] did not change the relative amounts of two fractions, though peak separation increased. MWD of HMWF was ~1.3 to ~1.5.

These data cannot be directly compared with those reported in Table 1 and 2, since the polarities of these systems were different. Thus, while for the temperature

Table 5. GPC analysis of molecular weights of PIB prepared using t-BuBr/Et$_2$AlCl/MeCl system at various monomer concentrations

[i-C$_4$H$_8$] M	Yield %	$\bar{M}_v{}^a$ ×10^{-3}	Total Polymer ×10^{-3}			L M W F ×10^{-3}			H M W F ×10^{-3}			HMWF %
			$\bar{M}_n$	$\bar{M}_w$	MWD	$\bar{M}_n$	$\bar{M}_w$	MWD	$\bar{M}_n$	$\bar{M}_w$	MWD	
−30 °C												
1.8	77	—	73	178	2.4$_0$	—	—	—	—	—	—	—
2.4	66	179	80	178	2.2$_3$	—	—	—	—	—	—	—
3.0	55	200	98	229	2.3$_3$	—	—	—	—	—	—	—
3.6	25	209	139	335	2.4$_1$	—	—	—	—	—	—	—
4.2	38	233	95	241	2.5$_5$	—	—	—	—	—	—	—
4.8	12	251	138	311	2.2$_6$	—	—	—	—	—	—	—
−40 °C												
1.8	25	—	97	269	2.7$_7$	—	—	—	—	—	—	—
2.4	40	—	136	336	2.4$_7$	—	—	—	—	—	—	—
3.0	30	—	100	254	2.5$_3$	—	—	—	—	—	—	—
3.6	12	—	171	346	2.0$_2$	—	—	—	—	—	—	—
4.2	21	—	122	326	2.6$_7$	—	—	—	—	—	—	—
4.8	11	—	195	521	2.6$_8$	—	—	—	—	—	—	—
−50 °C												
1.8	25	—	148	371	2.5$_0$	83	122	1.4$_6$	432	579	1.3$_4$	52
2.4	28	335	167	441	2.6$_4$	86	100	1.1$_7$	422	636	1.5$_1$	59
3.0	10	365	128	477	3.7$_3$	74	127	1.7$_2$	552	840	1.5$_0$	49
3.6	8	378	171	637	3.6$_8$	92	137	1.4$_9$	716	1059	1.4$_8$	55
4.2	13	458	155	652	4.2$_2$	90	211	2.3$_5$	788	1255	1.6$_0$	43

[Et$_2$AlCl] = 6.0 × 10^{-3} M, [t-BuBr] = 4.5 × 10^{-5} M, MeCl = 12 ml, cosolvent-n-pentane.
a Data from Table 6.

study (Tables 1 and 2) MeCl concentration was ~75 vol.% (15 ml), in the [M] study, they were 65 vol.% (13 ml, Table 4) and 60 vol.% (12 ml, Table 5). Qualitatively, however, the effects of increasing [M] and decreasing temperature are similar in that both changes result in an increase of $\bar{M}_n$ and $\bar{M}_w$ of the total polymer as well as HMWF. The MWD's of two fractions were unaffected by these changes in [M] and temperature. In contrast, decreasing temperature tended to decrease $\bar{M}_n$ and $\bar{M}_w$ of LMWF while increase in [M] tended to increase them. Also, temperature change did not affect the relative amounts of HMWF and LMWF, while an increase in [M] significantly increased HMWF at the expense of LMWF. The effect of [M] on the $\bar{M}_n$'s of the HMWF and LMWF is analyzed by means of Mayo plots in Section 4.

4. Elucidation of Relative Rate Constants Using GPC Data

This section concerns studies directed toward an elucidation of relative rate constants, k_{tr}/k_p and k_t/k_p, and a mechanistic interpretation of these parameters. The data for the HMWF of PIB samples prepared with t-BuCl and t-BuBr/Et$_2$AlCl/MeCl systems shown in Table 7 were calculated from Mayo plots (see also p. 56 for Mayo Equation) using $\bar{M}_n$ data given in Tables 4 and 5. The data of LMWF of PIB prepared with t-BuBr/Et$_2$AlCl/MeCl system at −50 °C shown in Table 7 were calculated from Mayo plots using $\bar{M}_n$ data given in Table 5.

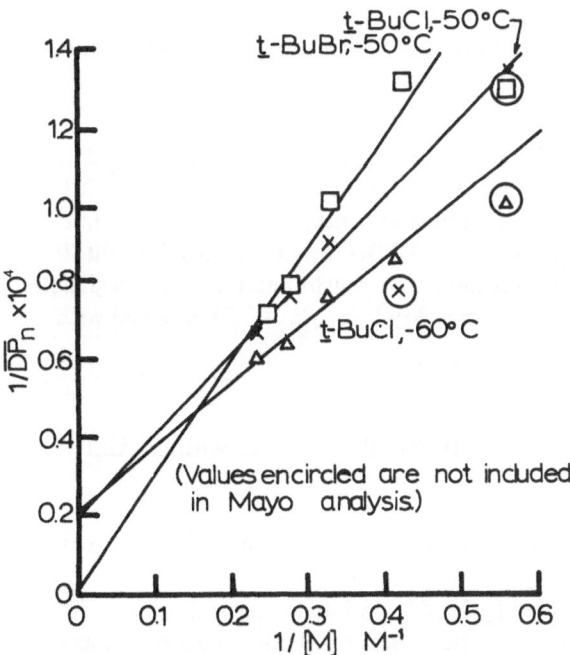

Fig. 7. Mayo plots based on GPC $\bar{M}_n$ values of PIB prepared with t-BuX/Et$_2$AlCl/MeCl at −50 °C and −60 °C

For the HMWF obtained with the t-BuCl/Et$_2$AlCl initiator system, k_t/k_p = 2.07 x 10^{-4} at −50 °C and 1.58 x 10^{-4} at −60 °C, while k_{tr}/k_p = 1.91 x 10^{-5} ≅ 0, and 2.14 x 10^{-5} ≅ 0, respectively (Fig. 7). The increase in molecular weight with decreasing temperature is thus due to a decrease in k_t/k_p. The k_t/k_p of the LMWF could not be determined because of insufficient slope definition in the Mayo plot.

The Mayo plot for the HMWF of PIB prepared with t-BuBr/Et$_2$AlCl/MeCl system at −50 °C gives k_t/k_p = 3.58 x 10^{-4} and k_{tr}/k_p = 0 (Fig. 7). For the LMWF of the same PIB samples, k_t/k_p = 1.95 x 10^{-4} and k_{tr}/k_p = 5.67 x 10^{-4} indicating transfer and termination are comparable for the LMWF at −50 °C (Fig. 8).

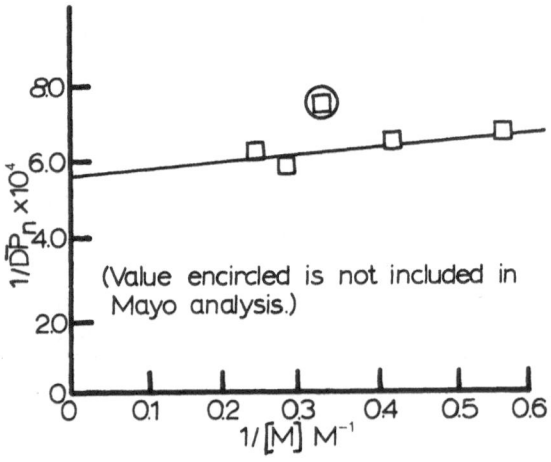

Fig. 8. Mayo plot based on GPC $\bar{M}_n$ values of LMWF of PIB prepared with t-BuBr/Et$_2$AlCl/MeCl at −50 °C

These relative rate constants help illuminate the reaction mechanism and molecular weight controlling events. For the HMWF obtained with the t-BuCl/Et$_2$AlCl system at −50° and −60 °C, transfer to monomer is negligible and termination is molecular weight governing. Similarly, for the HMWF obtained with the t-BuCl/Et$_2$AlCl at −50 °C termination controls molecular weight and transfer is negligible. However, transfer is more important for the LMWF (k_{tr}/k_t = 2.8) obtained with t-BuBr/Et$_2$AlCl system.

5. Mechanistic Interpretation of Bimodal MWD of PIB Prepared with Et$_2$AlCl Coinitiator System

An interpretation of GPC data presented in Sections 2 and 3 must be consistent with the following facts:

1) GPC analysis of PIB's prepared by Et$_2$AlCl coinitiator at and below −50 °C shows bimodal MWD (Figs. 1 and 2) and the presence of bimodal distributions is also evident at different monomer concentrations (Figs. 5 and 6). In contrast, PIB's produced by Et$_2$AlBr and Et$_2$AlI show monomodal distributions (Figs. 3 and 4).

2) Temperature does not affect the relative amounts of LMWF and HMWF (Tables 1 and 2).

3) A decrease in temperature tends to increase the $\overline{M}_n$ and $\overline{M}_w$ of HMWF and to decrease $\overline{M}_n$ and $\overline{M}_w$ of LMWF (Tables 1 and 2).

4) Increasing monomer concentration increases the relative amounts of the HMWF obtained with t-BuCl initiator and merely increases the GPC peak separation without affecting the relative amounts of the LMWF and HMWF obtained with t-BuBr (Tables 4 and 5). Increasing monomer concentration tends to increase $\overline{M}_n$ and $\overline{M}_w$ of the HMWF and LMWF (Tables 4 and 5).

At least three possibilities may give rise to two different but simultaneous mechanisms producing bimodal MWD: the simultaneous propagation by two different, non-equilibrating chain carriers, e.g., free-ions and ion-pairs; the presence of two different counteranions, and the presence of an impurity, e.g. water.

The above facts can only be explained by assuming the presence of an impurity (e.g., water) in the system. Since water reacts slowly with Et_2AlCl at low temperatures, it may survive and function as a chain transfer and/or terminating agent leading to the formation of LMWF. The molecular weight of the HMWF is determined by termination, i.e., alkylation of the growing chain by the counteranion. The absence of bimodality in samples prepared using Et_2AlBr and Et_2AlI coinitiator systems may be due to rapid scavening of water by these relatively more reactive coinitiators.

The presence of water would also explain the absence of a significant temperature effect on the relative amounts of HMWF and LMWF. Decreasing temperatures, when the temperature is already way below the freezing point of water, has no effect on its concentration and so the relative amount of LMWF does not change (Tables 1 and 2).

The increase in $\overline{M}_n$ and $\overline{M}_w$ of the HMWF and the decrease in $\overline{M}_n$ and $\overline{M}_w$ of the LMWF with a decrease in temperature (Tables 1 and 2) also suggest H_2O to be a cause of bimodal MWD. Since the rate of initiation decreases with decreasing temperatures leading to a decrease in initiating carbenium ion concentration, the poisoning is more severe and hence the $\overline{M}_n$ and $\overline{M}_w$ of LMWF decrease. On the other hand, termination by alkylation by counteranions is reduced as the temperature is lowered and hence $\overline{M}_n$ and $\overline{M}_w$ of the HMWF increase.

An increase in [M] leads to an increase in the amount of HMWF using t-BuCl/Et_2AlCl system. Since the rate of initiation increases with an increase in [M], the scavenging of the impurity accelerates too, leading to a decrease in the amount of LMWF. Since initiation with t-BuBr is slower than with t-BuCl, scavenging must also be slower, and hence the concentration of impurity does not significantly decrease. Consequently, the amount of LMWF remains unaffected by an increase in [M].

An increase in [M] was also found to increase $\overline{M}_n$ and $\overline{M}_w$ of both fractions. This may be due to an increase in the rate of propagation while the rates of molecular weight controlling events remained unaffected.

Assumptions invoking two types of counteranions or chain carriers, e.g., free-ions and ion-pairs, would not explain the bimodal MWD's obtained with Et_2AlCl and monomodal MWD's with Et_2AlBr and Et_2AlI, or the effect of temperature on the relative amounts of two fractions. Decreasing temperatures would increase the dielectric constant of the medium and thereby the free-ion concentration, which in turn would lead to a change in the relative amounts of two fractions. Similarly, the effect of [M] on the relative amounts of two fractions could not be explained assuming two kinds of counteranions, since in the presence of two kinds of counteranions the relative amounts of HMWF and LMWF would remain constant. Also, the decrease in $\bar{M}_n$ and $\bar{M}_w$ of LMWF (or near constancy of them for t-BuBr/Et$_2$AlCl system) with a decrease in temperature could not be explained by either of these two assumptions both of which would predict an increase in $\bar{M}_n$ and $\bar{M}_w$.

The assumption that an impurity, probably water, gives rise to bimodal MWD with t-BuX/Et$_2$AlCl system was confirmed directly by experiments. First polymerizations were carried out at -60 °C by using the "H$_2$O"/Et$_2$AlCl system, $i.e.$, by increasing the moisture content in the enclosure from the usual <30 ppm to ~150 ppm. Yields were ~30% or less. The GPC trace of these PIB's (Fig. 9) coincides with those

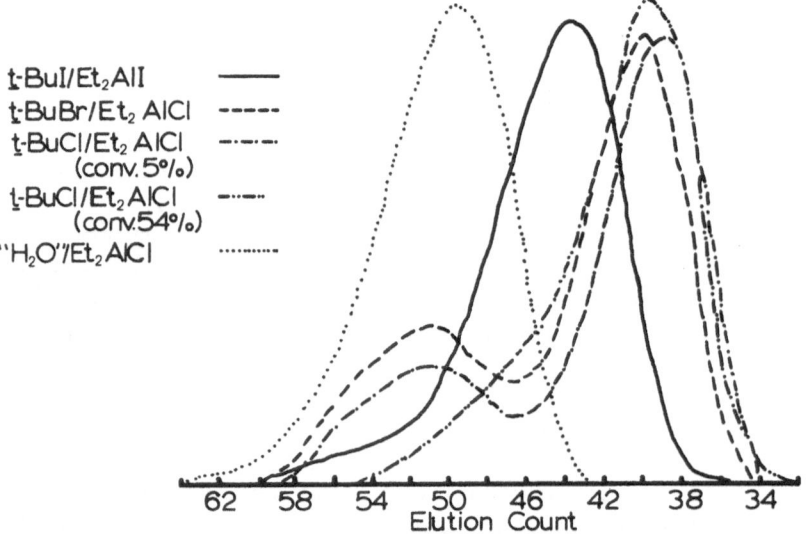

Fig. 9. GPC curves of PIB prepared using MeCl solvent at -60 °C

of the LMWF prepared with t-BuX/Et$_2$AlCl at -60 °C. Importantly, PIB's prepared with "H$_2$O"/Et$_2$AlCl did not contain HMWF. Evidently the polymerization was initiated by excess water and the molecular weights were also governed by chain transfer and/or termination due to excess water.

The following experiment further corroborates the proposition that bimodal MWD's are caused by "H$_2$O". Isobutylene polymerizations were carried out with t-BuCl/Et$_2$AlCl/MeCl at -60 °C to 5 and 54% conversions. According to the GPC traces of the products (Fig. 9), the sample obtained at higher conversion contains no LMWF. Since the impurity is increasingly scavenged with increasing conversion, the amount of LMWF would also be expected to decrease. Such a decrease in LMWF would not occur in the presence of two types of counteranions or growing species of various ionicities.

The following facts emerge based on the GPC study and reactivity studies[1].

1) PIB's prepared with the t-BuCl or t-BuBr/Et$_2$AlCl systems show bimodal distributions and polymerization does not occur with Et$_2$AlCl in the absence of t-BuX,

2) PIB's prepared with t-BuCl/Et$_2$AlBr have monomodal MWD and polymerization does not occur with Et$_2$AlBr in the absence of added t-BuX, and

3) PIB's prepared using t-BuI/Et$_2$AlI show monomodal distribution but slow polymerization occurs even in the absence of purposefully added t-BuX.

These results may be explained by the following scheme:

$$H_2O + Et_2AlX \underset{K}{\rightleftharpoons} [H^{\oplus}Et_2AlXOH^{\ominus}]$$

$$[H^{\oplus}Et_2AlXOH^{\ominus}] \begin{array}{c} \xrightarrow{R_1} EtH + EtAlXOH \quad \text{Termination} \\ \underset{M \quad R_2}{\xrightarrow{\hspace{2cm}}} HM^{\oplus}Et_2AlXOH^{\ominus} \longrightarrow \text{Propagation} \end{array}$$

For Et$_2$AlI system, both K and R$_2$ are relatively high and traces of water are consumed by initiation. For Et$_2$AlBr, K and R$_1$ are relatively large leading to fast scavenging of water. However, for Et$_2$AlCl systems, K is relatively small so that water may survive to function as a chain transfer and/or terminating agent.

In sum, the PIB's prepared using the t-BuX/Et$_2$AlX/MeX systems at different temperatures and monomer concentrations have been analyzed by GPC. While bimodal MWD was obtained with Et$_2$AlCl-based systems, monomodal distributions were found with Et$_2$AlBr and Et$_2$AlI coinitiators. The $\overline{M}_n$, $\overline{M}_w$ and MWD of total polymer, HMWF and LMWF were determined. The effect of temperature on the MWD of the total polymer is greater for samples obtained with Et$_2$AlCl coinitiator than with Et$_2$AlBr or Et$_2$AlI. For the HMWF of PIB's prepared with t-BuCl/Et$_2$AlCl and for PIB's prepared with Et$_2$AlI systems $\Delta E_{\overline{M}_n}$'s and $\Delta E_{\overline{M}_w}$'s calculated from GPC data were, within experimental error, equal to $\Delta E_{\overline{M}_v}$'s determined from $\overline{M}_v$ data, i.e., -1.8 kcal/mole for t-BuCl/Et$_2$AlCl and -4.6 kcal/mole for t-BuI/Et$_2$AlI (Table 8). The effect of [M] on $\overline{M}_n$ was determined and k_t/k_p and k_{tr}/k_p were calculated for the HMWF and LMWF obtained with t-BuCl and t-BuBr/Et$_2$AlCl/MeCl systems at $-50°$ and $-60°C$. It is postulated that the bimodal MWD obtained for the samples prepared by t-BuX/Et$_2$AlCl/MeCl system is due to the presence of an impurity, probably water.

IV. Viscosity Average Molecular Weights of Polyisobutylenes

1. Introduction

Among the various kinds of molecular weights, $\overline{M}_n$'s give direct, readily interpretable information in regard to the mechanism of elementary events like chain transfer or termination, however, their determination may be time consuming and relatively costly. In contrast, $\overline{M}_v$'s can be determined readily, inexpensively and with great precision[14]. Historically, M_v's have been used in isobutylene polymerization mechanism studies since ~1953, when Flory[15] first used $\bar{M}_v$ data to calculate $\Delta E_{\bar{M}_v}$ for the "H$_2$O"/BF$_3$ initiator system. Laster, Plesch[16], Kennedy and Thomas[13] and subsequently Kennedy and Squires[17] used $\bar{M}_v$ and $\Delta E_{\bar{M}_v}$ data to propose mechanism of isobutylene polymerization. Imanishi and co-workers[12, 18] have used $\bar{M}_v$ data extensively to calculate relative rate constants, k_t/k_p and k_{tr}/k_p and compared the proposed mechanisms of isobutylene, styrene and α-methylstyrene polymerizations.

Since the early 50's a large body of $\bar{M}_v$ and $\Delta E_{\bar{M}_v}$ data have been accumulated by many authors, including Plesch[19], Marek[11], and Kennedy[4]. Table 8 gives a comprehensive list of $\Delta E_{\bar{M}_v}$'s published as well as those determined in this work.

The validity of $\bar{M}_v$ in mechanism studies was further established by Kennedy, Williams and Shinkawa[10], who compared the molecular weight of PIB, obtained with γ-ray induced polymerization or Friedel-Crafts halides using osmometry, GPC and viscometry. The Arrhenius plots obtained with all these molecular weight data were parallel to each other, indicating a fairly constant activation energy difference of molecular weights, i.e., $\Delta E_{\bar{M}_n} = \Delta E_{\bar{M}_v} = \Delta E_{\bar{M}_w}$.

Kennedy and Rengachary[4] similarly determined $\Delta E_{\bar{M}_v}$, $\bar{M}_v$ and $\bar{M}_n$ for PIB prepared by a host of alkylaluminum initiator systems. Recently, Cesca and co-workers[20] have used $\bar{M}_v$ data to determine the mechanism of isobutylene polymerization using the Cl_2/Et$_2$AlCl/MeCl system.

Since our aim was to investigate in detail the mechanism of isobutylene polymerization by t-BuX/Et$_2$AlX/MeX systems, the acquisition of a large number of molecular weight data was essential. In view of the above considerations and to facilitate the comparison of our data with $\bar{M}_v$ data of previous workers, $\bar{M}_n$ studies reported earlier[1] have been extended by $\bar{M}_v$ determinations and are discussed in this section. The GPC study has indicated that MWD for PIB's prepared with t-BuCl/Et$_2$AlBr/MeBr and t-BuI/Et$_2$AlI/MeCl is invariably between ~1.7 to ~2.3 and ~2 to ~3, respectively. The MWD of HMWF obtained with t-BuX/Et$_2$AlCl/MeCl is from ~1.5 to ~2.0. These findings suggest that for a given initiator system, $\bar{M}_v/\bar{M}_n$ is constant and hence $\bar{M}_v$ can be used in mechanism studies. Also, for t-BuCl/Et$_2$AlCl/MeCl $\Delta E_{\bar{M}_n} = \Delta E_{\bar{M}_w} = \Delta E_{\bar{M}_v} = -1.8$ kcal/mole for the HMWF, and for the t-BuI/Et$_2$AlI/MeCl system $\Delta E_{\bar{M}_n} = \Delta E_{\bar{M}_w} = \Delta E_{\bar{M}_v} = -4.6$ kcal/mole (Table 8). These findings further justify use of $\bar{M}_v$ data in mechanism studies.

2. Effect of t-BuX Initiator, Et$_2$AlX Coinitiator, MeX Solvent and Temperature on $\bar{M}_v$ and $\Delta E_{\bar{M}_v}$ of PIB

The present study, parts of which have been reported earlier[21], concerns an analysis of $\bar{M}_v$ and $\Delta E_{\bar{M}_v}$ data obtained in isobutylene polymerization using the t-BuX/Et$_2$AlX/MeX and "H$_2$O"/EtAlCl$_2$/n-pentane systems. This section is devoted to a description of experimental findings, whereas their comprehensive discussion is deferred to Sections V and VI.

$\bar{M}_v$ data obtained using t-BuX/Et$_2$AlCl/MeCl from $-30°$ to $-70\,°C$ shown in Table 4 of the previous publication[1], have been used to construct the log $\bar{M}_v$ vs 1/T

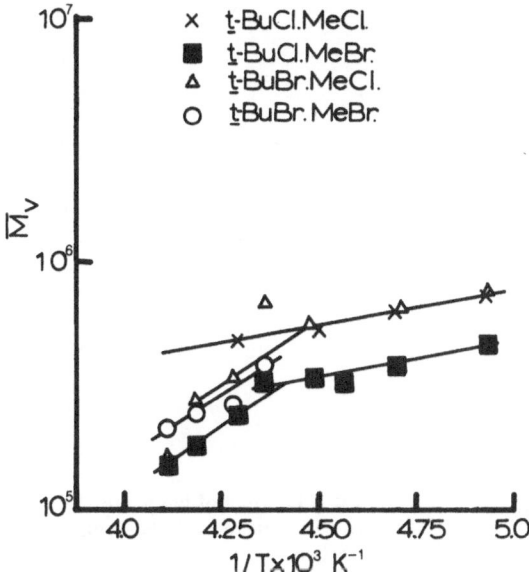

Fig. 10. Effect of temperature on $\bar{M}_v$ of PIB prepared with Et_2AlCl co-initiator

(Arrhenius) plots in Fig. 10. $\bar{M}_v$'s increased with a decrease in temperature. The highest and lowest $\bar{M}_v$'s were obtained with the t-BuCl/Et_2AlCl/MeCl and t-BuCl/ Et_2AlCl/MeBr systems, respectively. The $\bar{M}_v$'s produced using t-BuCl and t-BuBr in MeCl were indistinguishable in the $-50°$ to $-70\,°C$ range. However, lower $\bar{M}_v$'s were obtained with t-BuBr than t-BuCl from $-30°$ to $-45\,°C$.

For t-BuCl/Et_2AlCl/MeCl, $\Delta E_{\bar{M}_v} = -1.8$ kcal/mole in the range from $-30°$ to $-70\,°C$ and for t-BuCl/Et_2AlCl/MeBr, $\Delta E_{\bar{M}_v} = -4.6$ and -1.8 kcal/mole from $-30°$ to $-45\,°C$ and $-50°$ to $-70\,°C$, respectively. Similarly, for t-BuBr/Et_2AlCl/MeCl, $\Delta E_{\bar{M}_v} = -4.6$ and -1.8 kcal/mole, from $-30°$ to $-50\,°C$ and $-50°$ to $-70°$, respectively; with t-BuBr/Et_2AlCl/MeBr, $\Delta E_{\bar{M}_v} = -4.6$ kcal/mole, from $-30°$ to $-45\,°C$. These data are reproducible to ± 1.0 kcal/mole and are listed in Table 8 along with those reported in literature.

Table 5 of the previous publication[1] and Fig. 11 give $\bar{M}_v$'s obtained using the t-BuX/Et_2AlBr/MeX system. The effect of changing halogen in t-BuX or MeX (X = Cl, Br) on $\bar{M}_v$ is small. $\bar{M}_v$'s increase with a decrease in temperatures from $-25\,°C$ to $-65\,°C$.

For t-BuCl/Et_2AlBr/MeCl, $\Delta E_{\bar{M}_v} = -4.6$ and -1.8 kcal/mole in the range from $-30°$ to $-50\,°C$ and $-50°$ to $-70\,°C$, respectively. However, for the other three systems, $\Delta E_{\bar{M}_v} = -4.6$ kcal/mole over the entire temperature range.

While changing the solvent from MeCl to MeBr significantly reduced $\bar{M}_v$'s obtained with t-BuCl/Et_2AlCl, this decrease was small for t-BuCl/Et_2AlBr and t-BuBr/Et_2AlCl and was virtually absent for the t-BuBr/Et_2AlBr system. The magnitude of this effect seems to follow the decrease in nucleophilicities of counteranions: $Et_2AlCl_2^{\ominus} > Et_2AlClBr^{\ominus} > Et_2AlBr_2^{\ominus}$.

Table 6 of the preceding publication[1] and Fig. 12 record $\bar{M}_v$'s obtained using Et_2AlI coinitiator and MeCl solvent. PIB could not be obtained using MeBr or MeI. Highest $\bar{M}_v$'s were obtained in the absence of t-BuX initiator, $i.e.$, initiation by

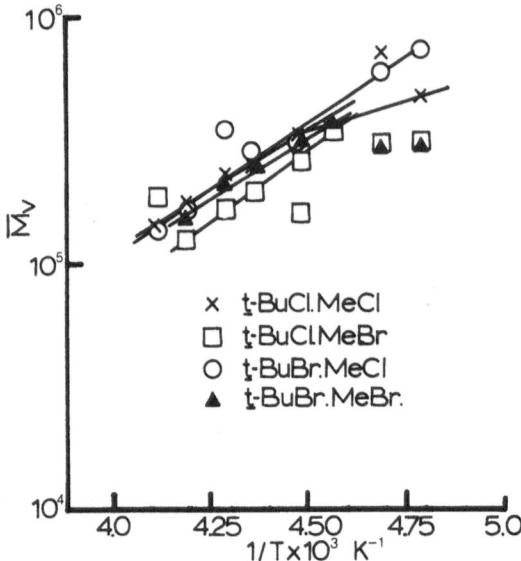

Fig. 11. Effect of temperature on $\overline{M}_v$ of PIB prepared with Et_2AlBr co-initiator

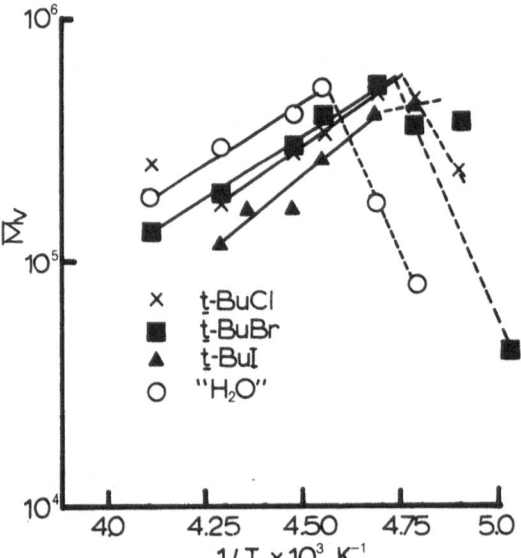

Fig. 12. Effect of temperature on $\overline{M}_v$ of PIB prepared with Et_2AlI co-initiator

"H_2O". While lowest $\overline{M}_v$'s were formed with t-BuI, $\overline{M}_v$'s obtained with t-BuCl and t-BuBr were similar and fell between those produced by "H_2O" and t-BuI.

$\overline{M}_v$'s increased with decrease in temperature in the range from $-30°$ to $-60\ °C$; below a certain temperature $\overline{M}_v$'s suddenly dropped with decreasing temperatures. The $\Delta E_{\overline{M}_v}$'s calculated for the higher temperature range $= -4.6$ kcal/mole for

Table 6. Effect of monomer concentration on $\overline{M}_V$ of PIB

Initiator Temp.	[Iso-butylene] M	$\dfrac{1}{[\text{Iso-butylene}]}$ M^{-1}	Yield[a] %	$\overline{M}_V$[a] $\times 10^{-3}$	$1/\overline{DP}_V$ $\times 10^4$	k_t/k_p	k_{tr}/k_p	k_t/k_{tr}
t-BuCl	1.8	0.56	7.5	256	2.19			
−40 °C	2.4	0.42	14.3	364	1.54	$3.75 \times$	0	−
	3.0	0.33	18.0	468	1.20	10^{-4}		
	3.6	0.28	22.1	470	1.19			
	4.2	0.24	30.1	597	0.94			
t-BuCl	1.8	0.56	40.0	332	1.69			
−50 °C	2.4	0.42	21.0	436	1.29	$3.21 \times$	0	−
	3.0	0.33	21.0	563	0.99	10^{-4}		
	3.6	0.28	36.0	669	0.84			
	4.2	0.24	38.5	870	0.64			
t-BuCl	1.8	0.56	13.7	222	2.53			
−60 °C	2.4	0.42	31.3	604	0.93	$2.49 \times$	0	−
	3.0	0.33	26.7	640	0.88	10^{-4}		
	3.6	0.28	24.9	1064	0.52			
	4.2	0.24	26.3	1209	0.46			
t-BuBr	2.4	0.42	36.2	179	3.14			
−30 °C	3.0	0.33	42.3	200	2.80	$4.20 \times$	$1.41 \times$	2.9_8
	3.6	0.28	28.9	209	2.69	10^{-4}	10^{-4}	
	4.2	0.24	18.3	233	2.40			
	4.8	0.21	14.0	251	2.24			
t-BuBr	2.4	0.42	19.9	335	1.67			
−50 °C	3.0	0.33	10.3	365	1.54	$3.10 \times$	$5.00 \times$	6.2_0
	3.6	0.28	4.7	378	1.48	10^{-4}	10^{-5}	
	4.2	0.24	7.1	458	1.23			
	4.8	0.21	2.8	498	1.13			

[Et$_2$AlCl] = 6.0×10^{-3} M, [t-BuX] = 4.5×10^{-5} M, MeCl = 13 ml for t-BuCl, 12 ml for t-BuBr, n-pentane cosolvent.

[a] Averages of at least two data points.

t-BuCl (−30° to −60 °C), t-BuBr (−30° to −60 °C), t-BuI (−40° to −65 °C) and "H$_2$O" (−30° to −55 °C). Probable reasons for the sudden decrease in $\overline{M}_V$ with decreasing temperatures are discussed in Section VI.

$\overline{M}_V$'s obtained in polymerizations induced by "H$_2$O"/EtAlCl$_2$/n-pentane are reported in Table 7 of the previous publication[1] and Fig. 13. While $\overline{M}_V$'s increased rapidly in the −30° to −80 °C range, the increase was smaller from −80° to −102 °C. $\Delta E_{\overline{M}_V}$'s = −4.6 kcal/mole (−30° to −80°) and −0.7 kcal/mole (−80° to −102 °C).

In sum, the effect of the nature of the halogen in t-BuX, Et$_2$AlX, and MeX, where X = Cl, Br, I, and temperature on $\overline{M}_V$ has been investigated. The highest $\overline{M}_V$'s were obtained with the t-BuCl/Et$_2$AlCl/MeCl system while the lowest $\overline{M}_V$'s were obtained using t-BuI/Et$_2$AlI/MeCl. Experiments have also been carried out with the

Table 7. Comparative study of kinetic parameters obtained using Mayo plots of $\bar{M}_v$ and $\bar{M}_n$ of PIB

Expt. Conditions	Temp. °C	$k_t/k_p \times 10^4$	$k_{tr}/k_p \times 10^5$	k_t/k_{tr}
t-BuCl[a]	−40°	3.75	0	—
$\bar{M}_v$	−50°	3.21	0	—
	−60°	2.49	0	—
$\bar{M}_n$ HMWF	−50°	2.07	1.91	10.8
$\bar{M}_n$ HMWF	−60°	1.58	2.14	7.4
t-BuBr[a]	−30°	4.20	14.1	2.98
$\bar{M}_v$	−50°	3.10	5.00	6.20
$\bar{M}_n$ HMWF	−50°	3.58	0	—
$\bar{M}_n$ LMWF	−50°	1.94	56.7	0.34

[a] With $Et_2AlCl/MeCl/n$-Pentane.

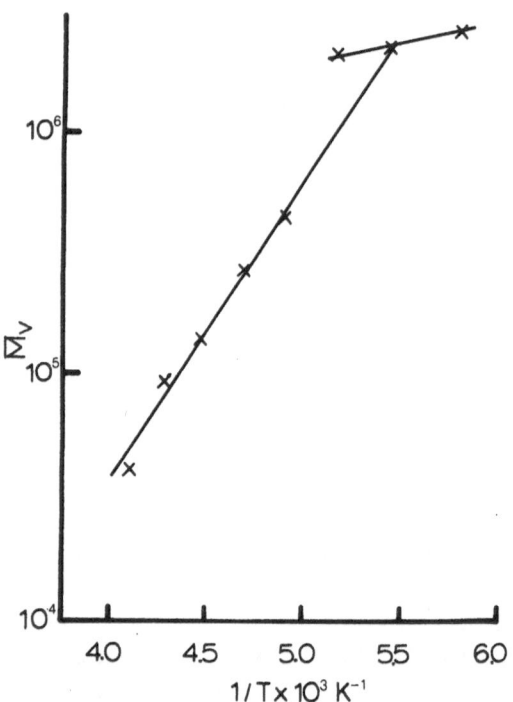

Fig. 13. Effect of temperature on $\bar{M}_v$ of PIB prepared with $H_2O/EtAlCl_2/$ n-pentane system

"H_2O"/$EtAlCl_2$/n-pentane system. Further, the effect of the nature of the initiator systems, solvents and temperature on $\Delta E_{\bar{M}_v}$ was also investigated. Table 8 lists $\Delta E_{\bar{M}_v}$'s obtained in this research together with similar, previously published data, along with the initiator systems used and temperature ranges. The effect of temperature and nature of initiator systems on $\bar{M}_v$ and $\Delta E_{\bar{M}_v}$ is discussed in Sections V and VI.

Table 8. $\Delta E_{\overline{M}_V}$'s as a function of initiator/coinitiator/solvent and temperature in isobutylene polymerization

1: $\underline{\Delta E}_{\overline{M}_V} = -6.6 \pm 1.0$ kcal/mole (-27.6 ± 4.2 kJ/mole)

No.	Initiator	Coinitiator	Solvent	Temp. Range, °C From	To	[M] M	$-\Delta E_{\overline{M}_V}$	Ref.
1	"H$_2$O"	AlCl$_3$	MeCl	−30°	−100°	3.1	5.6	17)
2	"H$_2$O"	AlCl$_3$	MeCl	−30°	−100°	3.1	6.6	10, 36)
3	"H$_2$O"	EtAlCl$_2$	MeCl	−30°	−100°	3.1	5.6	17)
4	"H$_2$O"	EtAlCl$_2$	MeCl	−30°	−100°	3.1	6.6	10, 36)
5	"H$_2$O"	BF$_3$	MeCl	−30°	−100°	3.1	5.6	17)
6	"H$_2$O"	BF$_3$	MeCl	−30°	−100°	3.1	6.6	10)
7	t-BuCl	Me$_2$AlCl	MeCl	−25°	−50°	3.1	7.0 ± 1.0	4)
8	t-BuBr	Me$_2$AlCl	MeCl	−25°	−50°	3.1	7.0 ± 1.0	4)
9	"H$_2$O"	MeAlCl$_2$	MeCl	−40°	−70°	3.1	7.0 ± 1.0	4)
10	H$_2$O	TiCl$_4$	CH$_2$Cl$_2$	+19°	−70°	0.4	8.2	16)
11	CCl$_3$COOH	TiCl$_4$	n-Hexane	−5°	−75°	1.25	7.5	19, 39)
12	H$_2$O	TiCl$_4$	EtCl	−30°	−112°	0.173	5.5 ± 0.5	19, 39)
13	H$_2$O	TiCl$_4$	EtBr	−38°	−103°	0.11−0.92	5.5 ± 0.5	19, 39)
14	"H$_2$O"	MgCl$_2$	n-Heptane	−10°	−78°	5.2	~6.0	40)
15	Cl$_2$	Et$_2$AlCl	MeCl	−30°	−50°	3.9	5.8 ± 0.4	41)
16	γ-rays	—	Bulk	+29°	−78°	12.4	6.3 ± 0.5	36)

2: $\underline{\Delta E}_{\overline{M}_V} = -4.6 \pm 1.0$ kcal/mole (-19.2 ± 4.2 kJ/mole)

No.	Initiator	Coinitiator	Solvent	Temp. Range, °C From	To	[M] M	$-\Delta E_{\overline{M}_V}$	Ref.
17	t-BuCl	Et$_2$AlCl	MeBr	−30°	−45°	3.0	4.6 ± 1.0	This work
18	t-BuBr	Et$_2$AlCl	MeCl	−30°	−50°	3.0	4.6 ± 1.0	This work
19	t-BuBr	Et$_2$AlCl	MeBr	−30°	−45°	3.0	4.6 ± 1.0	This work
20	t-BuCl	Et$_2$AlBr	MeCl	−30°	−50°	3.0	4.6 ± 1.0	This work
21	t-BuCl	Et$_2$AlBr	MeBr	−30°	−55°	3.0	4.6 ± 1.0	This work
22	t-BuBr	Et$_2$AlBr	MeCl	−30°	−65°	3.0	4.6 ± 1.0	This work
23	t-BuBr	Et$_2$AlBr	MeBr	−30°	−55°	3.0	4.6 ± 1.0	This work

Table 8 (continued)

2: $\underline{\Delta E}\bar{M}_V = -4.6 \pm 1.0$ kcal/mole (-19.2 ± 4.2 kJ/mole)

No.	Initiator	Coinitiator	Solvent	Temp. Range, °C		[M] M	$-\underline{\Delta E}\bar{M}_V$	Ref.
				From	To			
24	t-BuCl	Et₂AlI	MeCl	-30°	-60°	3.0	4.6 ± 1.0	This work
25	t-BuBr	Et₂AlI	MeCl	-30°	-60°	3.0	4.6 ± 1.0	This work
26	t-BuI	Et₂AlI	MeCl	-40°	-65°	3.0	4.6 ± 1.0	This work
27	"H₂O"	Et₂AlI	MeCl	-30°	-55°	3.0	4.6 ± 1.0	This work
28	"H₂O"	EtAlCl₂	n-Pentane	-30°	-80°	3.0	4.6 ± 1.0	11)
29	—	EtAlCl₂	n-Heptane	+21°	-55°	0.15	5.8	4)
30	t-BuBr	Et₂AlCl	MeCl	-25°	-60°	3.1	4.0 ± 0.5	4)
31	t-BuBr	Et₃Al	MeCl	-25°	-50°	3.1	4.0 ± 0.5	4)
32	—	AlBr₃	n-Heptane	+20°	-60°	0.05	3.9	42)
33	CF₃COOH	TiCl₄	n-Hexane	-60°	-80°	0.94	3.5 ± 1.5	19, 39)
34	"H₂O"	BF₃	Bulk	-10°	-110°	12.47	4.6	37, 15)
35	"H₂O"	AlCl₃	Propane	-50°	-100°	1.3	3.5	25)
36	"H₂O"	BF₃	CH₂Cl₂	-30°	-70°	1.45	3.3 ± 0.7	28)
37	hν	VCl₄	n-heptane	-20°	-78°	1.2	5.0	29)

3: $\Delta E\bar{M}_V = -1.8 \pm 1.0$ kcal/mole (7.5 ± 4.2 kJ/mole)

No.	Initiator	Coinitiator	Solvent	From	To	[M] M	$-\underline{\Delta E}\bar{M}_V$	Ref.
38	t-BuCl	Et₂AlCl	MeCl	-30°	-70°	3.0	1.8	This work
39	t-BuCl	Et₂AlCl	MeCl	-30°	-100°	3.1	1.7	36)
40	t-BuCl	Et₂AlCl	MeCl	-40°	-80°	3.1	2.0 ± 0.5	4)
41	t-BuCl	Et₂AlCl	MeCl	-40°	-100°	3.1	1.9	43)
42	Cl₂	Et₂AlCl	MeCl	-40°	-100°	3.1	1.9	43)
43	Cl₂	Et₂AlCl	MeCl	-35°	-75°	0.1	2.9	20)
44	Br₂	Et₂AlCl	MeCl	-30°	-75°	3.1	1.9	43)
45	t-BuBr	Et₂AlCl	MeCl	-45°	-60°	3.0	1.8	This work
46	t-BuCl	Et₂AlBr	MeCl	-45°	-60°	3.0	1.8	This work

47	I$_2$	Et$_2$AlI	CH$_2$Cl$_2$	−20°	−60°	4.36	2.8	44)
48	t-BuCl	Me$_2$AlCl	MeCl	−50°	−70°	3.1	2.0 ± 0.5	4)a
49	t-BuBr	Me$_2$AlCl	MeCl	−50°	−70°	3.1	2.0 ± 0.5	4)
50	"H$_2$O"	MeAlCl$_2$	MeCl	−70°	−100°	3.1	2.0 ± 0.5	4)
51	t-BuCl	i-Bu$_3$Al	MeCl	−30°	−100°	3.1	1.7	36)
52	t-BuCl	Et$_3$Al	MeCl	−40°	−80°	3.1	2.0	4)
53	t-BuCl	Et$_3$Al	MeCl	−50°	−100°	3.1	1.7	36)
54	t-BuCl	Me$_3$Al	MeCl	−20°	−78°	3.1	1.7	36)
55	t-BuCl	Me$_3$Al	MeCl	−40°	−100°	3.1	1.9	43)
56	Cl$_2$	Me$_3$Al	MeCl	−40°	−100°	3.1	1.9	43)
57	H$_2$O	SnCl$_4$	EtCl	−63.5	−95.5	3.2	0 ± 2	19)
58	H$_2$O	TiCl$_4$	n-Hexane	−60°	−80°	0.94	3 ± 1.5	19, 39)
59	"H$_2$O"	EtAlCl$_2$	n-Pentane	−80°	−102°	3.1	0.7	This work
60	"H$_2$O"	AlCl$_3$	Propane	−110°	−145°	1.3	0.22	25)
61	"H$_2$O"	AlCl$_3$	MeCl	−100°	−145°	3.1	0.7	17)
62	"H$_2$O"	BF$_3$	MeCl	−100°	−145°	3.1	0.7	17)
63	"H$_2$O"	EtAlCl$_2$	MeCl	−100°	−145°	3.1	0.7	17)
64	"H$_2$O"	BCl$_3$	CH$_2$Cl	−30°	−80°	1.04	1.8 ± 0.4	28)
65	t-BuCl	(CH$_2$=CH)$_3$Al	CH$_2$Cl$_2$	−30°	−70°	3.0	2.3 ± 0.28	45)
66	t-BuCl	Et$_2$AlCl	MeCl	−35°	−70°	0.1	2.9	24)

a $\Delta E \overline{M}_v$ for isobutylene-isoprene copolymerization.

3. Analysis of Molecular Weight Controlling Events Using Mayo Plots

The effect of monomer concentration [M] on the molecular weight may be expressed by the Mayo equation $\dfrac{1}{DP} = \dfrac{k_{tr}}{k_p} + \dfrac{k_t}{k_p} \cdot \dfrac{1}{[M]}$ and the corresponding first order Mayo plot provides relative rate constants of the molecular weight controlling events[22]. Transfer and/or termination with solvent, polymer and impurities are assumed to be negligible. Although the Mayo equation is strictly valid only for data obtained at low conversion, Plesch[23] has shown that the plot provides remarkably reliable information even with data obtained at relatively high conversions.

The Mayo equation has already proven its usefulness in elucidating the mechanism of isobutylene polymerization. Thus, Imanishi[12, 18] determined k_t/k_p and k_{tr}/k_p, using "H_2O"/$TiCl_4$, $TiCl_4 \cdot CCl_3COOH$ and $SnCl_4 \cdot CCl_3COOH$ initiators, n-C_7H_{14}, $CHCl_3$ and CH_2Cl_2 solvents at $-20°$, $-50°$ and $-78\,°C$. Marek[11] obtained k_t/k_p = 4.3×10^{-3} and k_{tr}/k_p = 2.8×10^{-2} for the $EtAlCl_2/n$-C_7H_{14} system at $-10\,°C$. According to Cesca and co-workers[20, 24] who used Cl_2 and t-$BuCl/Et_2AlCl/MeCl$ at $-45\,°C$, Mayo plots showed no intercept indicating k_{tr}/k_p = 0; k_t/k_p, as calculated by us from their data is $\sim 1.35 \times 10^{-4}$ and $\sim 9 \times 10^{-5}$, respectively.

The present study concerns the effect of isobutylene concentration on PIB $\bar{M}_v$ using t-$BuCl/Et_2AlCl/MeCl$ and t-$BuBr/Et_2AlCl/MeCl$ systems at various temperatures. Figures 14 and 15 and Table 6 contain the data obtained using the t-$BuCl$ initiator system. Significantly, the Mayo plots showed an absence of intercepts, i.e., $k_{tr}/k_p \simeq 0$, and k_t/k_p = 3.75×10^{-4}, 3.21×10^{-4} and 2.49×10^{-4} at $-40°$, $-50°$ and $-60\,°C$, respectively. The linear $\log k_t/k_p$ vs. $\dfrac{1}{T}$ (Arrhenius) plot (Fig. 15) gave $\Delta E_{k_t/k_p}$ = 2.4 kcal/mole. Data could not be obtained at $-30\,°C$ because of run-away polymerizations.

Figure 16 shows the Mayo plots obtained with the t-$BuBr/Et_2AlCl/MeCl$ system at $-30°$ and $-50\,°C$. These plots exhibit significant intercepts, i.e., k_{tr}/k_p = 1.4×10^{-4} and 0.5×10^{-4} and give k_t/k_p = 4.2×10^{-4} and 3.1×10^{-4} at $-30°$ and $-50\,°C$, respectively. Thus, both k_t/k_p and k_{tr}/k_p decrease with a decrease in temperature. Interestingly, k_t/k_{tr} increases from 3.0 at $-30\,°C$ to 6.2 at $-50\,°C$, which indicates that k_{tr} is more sensitive to changes in temperature than k_t.

These studies provide valuable insight into the molecular weight governing mechanisms operating in isobutylene polymerization using the t-$BuCl$ or t-$BuBr/$ $Et_2AlCl/MeCl$ systems. The absence of intercepts in the Mayo plots for the t-$BuCl/$ $Et_2AlCl/MeCl$ system indicates an absence of chain transfer to monomer. Evidently, termination controls molecular weights in this system. The absence of intercepts in the Mayo plots and the linearity of the $\log k_t/k_p$ vs. $1/T$ (Arrhenius) plot suggest molecular weight control by termination, at least in the range from $-40°$ to $-60\,°C$. The decrease in k_t/k_p with a decrease in temperature explains increase in molecular weights. The existence of termination also explains the incomplete conversions obtained with the t-$BuCl/Et_2AlCl/MeCl$ system[1].

For polymerizations initiated with the t-$BuBr/Et_2AlCl/MeCl$ system both transfer to monomer and termination are operative and their relative importance is temperature dependent. Evidently, transfer is less important than termination.

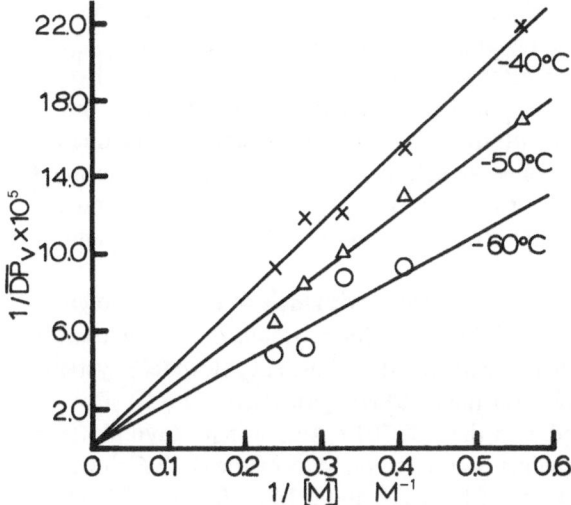

Fig. 14. Mayo plots based on $\overline{M}_v$'s of PIB's prepared with t-BuCl/Et$_2$AlCl/MeCl

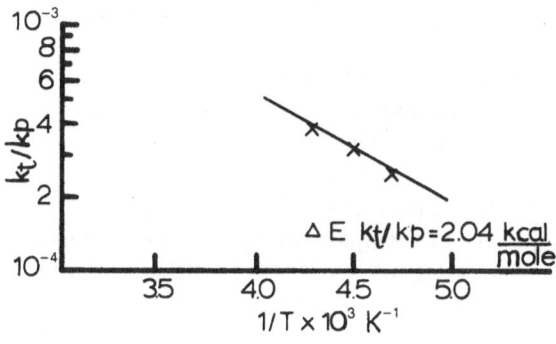

Fig. 15. Effect of temperature on k_t/k_p for isobutylene polymerization using t-BuCl/Et$_2$AlCl/MeCl system

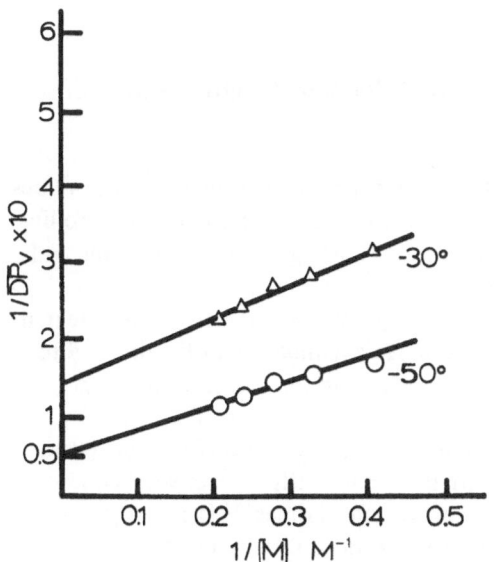

Fig. 16. Mayo plots based on $\overline{M}_v$'s of PIB's prepared with t-BuBr/Et$_2$AlCl/MeCl

Since increasing $\overline{M}_v$'s with decreasing temperatures are due to a decrease in k_t/k_p and/or k_{tr}/k_p, the corresponding $\Delta E_{\overline{M}_v}$'s reflect $\Delta E_{k_t/k_p}$ and/or $\Delta E_{k_{tr}/k_p}$. If either termination or monomer transfer alone control molecular weights $\Delta E_{\overline{M}_v}$ would be determined by either $\Delta E_{k_t/k_p}$ or $\Delta E_{h_{tr}/k_p}$. The similarity between the $\Delta E_{\overline{M}_v}$ and $\Delta E_{k_t/k_p}$ values obtained with the t-BuCl/Et$_2$AlCl system may be due to the fact that molecular weight control is exclusively by k_t/k_p (termination dominated polymerization). For the t-BuBr/Et$_2$AlCl system, both transfer and termination are molecular weight controlling and hence $\Delta E_{\overline{M}_v}$ is a function of both $\Delta E_{k_t/k_p}$ and $\Delta E_{k_{tr}/k_p}$.

Comparisons of relative rate constants obtained with $\overline{M}_v$'s of the total polymer and $\overline{M}_n$'s of the HMWF for the same samples show similar trends: negligible transfer and termination control of molecular weights for the t-BuCl/Et$_2$AlCl/MeCl system in the $-40°$ to $-60\,°C$ range and also for the t-BuBr/Et$_2$AlCl/MeCl at $-50\,°C$ (Table 7). For the samples prepared with the t-BuCl/Et$_2$AlCl system Mayo plots based on $\overline{M}_v$'s show zero intercept while that based on $\overline{M}_n$'s of the HMWF shows a small but finite intercept, $i.e.$, $k_{tr}/k_p = 1.91 \times 10^{-5}$ and 2.14×10^{-5} at $-50°$ and $-60\,°C$. Similarly, for the samples prepared with the t-BuBr/Et$_2$AlCl system the Mayo plot based on $\overline{M}_n$'s of HMWF shows zero intercept while the Mayo plot based on $\overline{M}_v$'s show a very small intercept, $i.e.$, $k_{tr}/k_p = 5.0 \times 10^{-5}$ at $-50\,°C$. The reasons for this small discrepancy are not known.

In sum, according to Mayo plot study, in isobutylene polymerization with t-BuCl/Et$_2$AlCl/MeCl system, termination controls molecular weights in the range from $-40°$ to $-60\,°C$, whereas with the t-BuBr/Et$_2$AlCl/MeCl system, molecular weight control is by monomer transfer and termination at $-30\,°C$, and predominately by termination at $-50\,°C$. These studies provide relative rate constants (k_t/k_p and k_{tr}/k_p) and help explain the effect of temperature on molecular weights. Finally, Mayo plots greatly helped substantiating the hypothesis according to which $\Delta E_{\overline{M}_v}$ values are diagnostic for the molecular weight governing mechanisms in isobutylene polymerization.

V. The Relationship Between $\Delta E_{\overline{M}_v}$ and Molecular Weight Control in Isobutylene Polymerization

In the course of these investigations it became apparent that the numerical values of $\Delta E_{\overline{M}_v}$'s contained important information concerning molecular weight controlling events. Thus, it was decided to search the literature thoroughly and to assemble and evaluate available information in this regard.

Table 8 is a comprehensive compilation of $\Delta E_{\overline{M}_v}$ values obtained in this research together with those available from the literature. An examination of the data led to a definition of the effect of initiating system, solvent and temperature, and to a general hypothesis on molecular weight control in isobutylene polymerization.

Previous workers[16, 17, 25] have often determined $\Delta E_{\overline{M}_v}$ values for isobutylene polymerization systems and tried to correlate these data with experimental conditions or fundamental mechanisms. Thus, Kennedy and Thomas[25] found $\Delta E_{\overline{M}_v} = -3.5$ kcal/mole and -0.22 kcal/mole in the range from $-50°$ to $-110\,°C$ and $-110°$

to $-145\,°C$, respectively, for the "H_2O"/$AlCl_3$/propane system. This change in $\Delta E_{\overline{M}_v}$ was explained by postulating that above $-110\,°C$ the viscosity of the medium was sufficiently low for fast monomer diffusion, whereas below $-110\,°C$ diffusion became molecular weight controlling. Kennedy and Squires[17] reported $\Delta E_{\overline{M}_v} = -6.6$ and -0.7 kcal/mole in the range from $-30°$ to $-80\,°C$ and $-100°$ to $-145\,°C$, respectively, for "H_2O"/$AlCl_3$, "H_2O"/BF_3 and "H_2O"/$EtAlCl_2$ using MeCl. The change in $\Delta E_{\overline{M}_v}$ was attributed to the change from transfer to solvent at higher temperatures to transfer to monomer of lower temperatures. Plesch[16] reported $\Delta E_{\overline{M}_v} = -8.2$ kcal/mole for H_2O/$TiCl_4$/CH_2Cl_2 in the range from $+19°$ to $-70\,°C$, whereas the slope of the Arrhenius plot decreased below $-70\,°C$, indicating a decrease in $\Delta E_{\overline{M}_v}$. Plesch explained these results by postulating that the growing species changed from ion-pairs to free-ions below $-70\,°C$. Analysis of the data in Table 8 shows that none of the above explanations are satisfactory.

A thorough examination of the data in Table 8 reveal that the $\Delta E_{\overline{M}_v}$ values fall into three groups, characterized by the following values (in kcal/mole): -6.6 ± 1.0, -4.6 ± 1.0 and -1.8 ± 1.0. (-27.6 ± 4.2, -19.2 ± 4.2 and -7.5 ± 4.2, in kJ/mole). The mechanistic significance of these three classes of $\Delta E_{\overline{M}_v}$ values becomes readily apparent by the following considerations:

For cationic olefin polymerization, in general, $E_{tr} > E_t > E_p$[19, 26, 27], where E_{tr}, E_t and E_p are the activation energies of chain transfer to monomer, termination and propagation. Since propagation is an ion-molecule reaction, it requires very little, if any, (0–2 kcal/mole) activation energy. E_t is also small (2–4 kcal/mole) since termination is a reaction between two oppositely charged species. Since transfer involves a substantial rearrangement of bonds, it requires the highest activation energy among the molecular weight controlling events. Assuming that termination by impurities, and transfer to solvent and polymer are negligible, $\Delta E_{\overline{M}_v} = E_p - E_t$ or $E_p - E_{tr}$ for systems with termination or transfer, respectively. In case both termination and transfer are important, $\Delta E_{\overline{M}_v}$ becomes a function of both E_t and E_{tr}. Since $E_{tr} > E_t$ and $E_{tr} - E_p > E_t - E_p$, it is postulated that the highest $\Delta E_{\overline{M}_v} = -6.6$ kcal/mole is characteristic for a system in which monomer transfer controls molecular weight while the lowest $\Delta E_{\overline{M}_v} = -1.8$ kcal/mole reflects molecular weight control by termination. For systems with $\Delta E_{\overline{M}_v} = -4.6$ kcal/mole, molecular weight control is by a combination of termination and chain transfer to monomer. A change of $\Delta E_{\overline{M}_v}$ with a change of initiator, solvent or temperature reflects a change in the molecular weight controlling mechanism.

According to this hypothesis, $\Delta E_{\overline{M}_v}$'s = -6.6 and -1.8 kcal/mole are extreme values, characteristic of two fundamentally different molecular weight governing processes and intermediate values between these extremes indicate the existence of a transition region in which both molecular weight controlling processes are operative. While a transition region between the extremes must always exist, its experimental definition may be difficult or even impossible, particularly when its operational range is narrow and/or insufficient data are available. In other words, the transition region may get "buried" in the respective regions for termination and/or transfer to monomer.

Unfavorable experimental conditions may altogether prevent the detection of termination or transfer region. For example, the transfer region for the t-BuCl/

$Et_2AlCl/MeCl$ system may lie above $-30\,°C$, *i.e.*, in a region where polymerization cannot be carried out, due to the boiling point of MeCl ($-24.2\,°C$), or the termination region for t-BuX/Et_2AlI/MeCl may lie below $-70\,°C$, where initiation does not take place.

According to the data in Table 8, termination is the dominant molecular weight determining event at low temperatures whereas chain transfer to monomer is molecular weight controlling at higher temperatures. This finding reflects the higher activation energy of transfer over that of termination. Evidently in many, if not all, isobutylene polymerization systems chain transfer may be selectively "frozen out" and transfer-less polymerizations can be readily obtained by judicious temperature control. Transfer-less polymerizations are of great synthetic value.

By raising the temperature a transfer-less isobutylene polymerization may be gradually converted into one in which molecular weights are predominantly controlled by chain transfer to monomer. Faster termination at higher temperatures should lead to lower yields, and indeed such a situation prevails with the "H_2O"/BCl_3/ CH_2Cl_2 system[28]. In instances where higher temperatures produce higher yields, conceivably, a rise in temperature increases the rate of initiation also compensating for rapid termination.

Another consequence of this hypothesis is that for every isobutylene polymerization system, there *must* exist three temperature regions over which molecular weight control is, respectively by termination, a combination of transfer to monomer and termination, and transfer to monomer alone; although due to experimental limitations all three regions may not be possible to detect.

According to the data obtained in this work, $\Delta E_{\overline{M}_v} = -1.8 \pm 1.0$ kcal/mole, is characteristic of the t-BuCl/Et_2AlCl/MeCl ($-30°$ to $-70°$), t-BuBr/Et_2AlCl/MeCl ($-50°$ to $-70°$) and t-BuCl/Et_2AlBr/MeCl ($-50°$ to $-70\,°C$) systems, and indicates molecular weight control by termination. For the rest of the initiator systems studied, $\Delta E_{\overline{M}_v} = -4.6 \pm 1.0$ kcal/mole, which suggests molecular weight control by a combination of termination and chain transfer to monomer. For the t-BuBr/ Et_2AlCl/MeCl system, a decrease in $\Delta E_{\overline{M}_v}$ from -4.6 to -1.8 kcal/mole with decreasing temperature reflects a change in molecular weight control from a combination of termination and transfer to monomer to predominantly termination.

For t-BuCl/Et_2AlCl/MeX a change in $\Delta E_{\overline{M}_v}$ from -1.8 to -4.6 kcal/mole brought about by changing the solvent from MeCl to MeBr suggests a change in molecular weight control from termination to a combination of termination and transfer to monomer.

This hypothesis is corroborated by results obtained from an analysis of Mayo plots, model experiments and consideration of findings of γ-ray induced isobutylene polymerizations. Thus, the effect of [M] on $\overline{M}_v$'s and $\overline{M}_n$'s was analyzed by Mayo plots to determine k_t/k_p, k_{tr}/k_p and k_t/k_{tr} at temperatures where, according to this hypothesis, molecular weight control is by termination, or a combination of termination and transfer to monomer. Thus, Mayo plots were constructed using data obtained with the t-BuCl/Et_2AlCl/MeCl system at $-40°$, $-50°$ and $-60\,°C$, *i.e.*, at temperatures where $\Delta E_{\overline{M}_v} = -1.8$ kcal/mole and therefore the molecular weights were expected to be termination controlled. Confirming this expectation, Mayo analyses using $\overline{M}_v$ data gave $k_{tr}/k_p \cong 0$, and $k_t/k_p = 3.74 \times 10^{-4}$, 3.21×10^{-4} and

2.49 x 10^{-4} at $-40°$, $-50°$ and $-60\,°C$, respectively. Similarly, Mayo plots using $\overline{M}_n$ (GPC) of HMWF gave negligibly low values for k_{tr} (i.e., $k_{tr}/k_p = 1.91 \times 10^{-5}$ and 2.41×10^{-5}) and $k_t/k_p = 2.07 \times 10^{-4}$ and 1.58×10^{-4} at $-50°$ and $-60\,°C$, respectively.

The Mayo plots for samples prepared with t-BuBr/Et$_2$AlCl/MeCl are similarly revealing. According to the Arrhenius plot for this system, $\Delta E_{\overline{M}_v} = -4.6$ kcal/mole in the $-30°$ to $-45\,°C$ range which suggests molecular weight control by a combination of termination and transfer to monomer, and $\Delta E_{\overline{M}_v} = -1.8$ kcal/mole in the $-50°$ to $-65\,°C$ range which indicates molecular weight control by termination alone. The Mayo plots using $\overline{M}_v$ at $-30\,°C$ gave $k_{tr}/k_p = 1.41 \times 10^{-4}$ and $k_t/k_p = 4.20 \times 10^{-4}$ and hence $k_t/k_{tr} = 2.98$ indicating comparable termination and transfer activity. At $-50\,°C$, the Mayo plots gave $k_{tr}/k_p = 0.50 \times 10^{-4}$, $k_t/k_p = 3.10 \times 10^{-4}$ and $k_t/k_{tr} = 6.20$, suggesting that termination becomes more important at lower temperatures. A similar conclusion is reached using Mayo plots of $\overline{M}_n$ data of the HMWF at $-50\,°C$, where $k_{tr}/k_p \cong 0$ and $k_t/k_p = 3.58 \times 10^{-4}$.

As shown above, for the "H$_2$O"/EtAlCl$_2$/n-pentane system, $\Delta E_{\overline{M}_v} = -4.6 \pm 1.0$ kcal/mole. Marek[11] obtained $\Delta E_{\overline{M}_v} = -5.8$ kcal/mole and $k_t/k_p = 4.3 \times 10^{-3}$ and $k_{tr}/k_p = 2.8 \times 10^{-2}$ at $-10\,°C$ for EtAlCl$_2$/n-heptane system. More recently, Marek[29] obtained $\Delta E_{\overline{DP}} = -5.0$ kcal/mole and $k_t/k_p = 5.40 \times 10^{-4}$ and $k_{tr}/k_p = 7.3 \times 10^{-4}$ for a hν/VCl$_4$/n-heptane system. Again the intermediate $\Delta E_{\overline{M}_v} = -4.6 \pm 1.0$ kcal/mole obtained reflects that the molecular weight governing event is a combination of transfer to monomer and termination and is confirmed by the ratio of rate constants.

Model experiments convincingly indicate molecular weight control by transfer for systems characterized by $\Delta E_{\overline{M}_v} = -6.6$ kcal/mole. Thus according to Kennedy and Rengachary[4] isobutylene polymerizations using the t-BuCl/Me$_2$AlCl/MeCl system gave $\Delta E_{\overline{M}_v} = -7.0$ kcal/mole in the range from $-25°$ to $-50\,°C$. Corresponding model experiments with 2,4,4-trimethyl-1-pentene and the same initiator system gave for the relative rates of elimination/alkylation = 30 ± 5, i.e., for the ratio corresponding to k_{tr}/k_t in polymerization. Evidently, and in line with this hypothesis, transfer was the major molecular weight controlling event.

Model experiments with 2,4,4-trimethyl-1-pentene (C$_8$H$_{16}$, TMP) and "H$_2$O"/AlBr$_3$/MeBr at $-80\,°C$, i.e., with a conventional Lewis acid system which would give $\Delta E_{\overline{M}_v} = -6.6$ kcal/mole in isobutylene polymerization, gave exclusively a dimer (C$_{16}$H$_{32}$) by proton elimination, i.e., by a mechanism which mimics transfer in polymerization:

$$C_8H_{16} + \text{"H}_2\text{O"}/\text{AlBr}_3 \xrightarrow{\text{MeBr}} H-C_8H_{16}^\oplus \ \text{AlBr}_3\text{OH}^\ominus$$

$$\Big\downarrow\ \text{TMP}\ (-H^\oplus\text{AlBr}_3\text{OH}^\ominus)$$

$$C_{16}H_{32}$$

Similar results were recently reported by Cesca and co-workers[30] using the "H$_2$O"/EtAlCl$_2$/MeCl system which is also characterized by $\Delta E_{\overline{M}_v} = -6.6$ kcal/mole (Table 8).

Importantly, also, isobutylene polymerizations initiated by γ-radiation in bulk are characterized by $\Delta E_{\bar{M}_v} = -6.6$ kcal/mole in the range from $+29°$ to -78 °C[52]. Since counteranion is absent in these systems, termination by counteranion must also be absent and molecular weight control can only occur by transfer to monomer.

In sum, a hypothesis suggesting a close relation between $\Delta E_{\bar{M}_v}$ and the molecular weight governing mechanisms has been developed. Experimentally readily obtainable $\Delta E_{\bar{M}_v}$ values have been found to fall into three groups, i.e. -6.6 ± 1.0, -4.6 ± 1.0 and -1.8 ± 1.0 kcal/mole which are proposed to reflect three different molecular weight governing mechanisms: transfer to monomer, a combination of transfer and termination, and termination, respectively. Experimental proofs in support of this hypothesis have been obtained by an analysis of Mayo plots and data provided by model experiments. Based on this work and on data published in the literature, a relation between the nature of counteranions and molecular weights in isobutylene polymerization has been proposed.

VI. Effect of Counteranion, Solvent and Temperature on Molecular Weight of Polyisobutylene

1. Effect of Counteranion on Molecular Weights in Isobutylene Polymerization

The effect of counteranion, $G^{\ominus}$, in cationic olefin polymerization has been studied in a variety of ways. Thus, effect of $G^{\ominus}$ in isomerization polymerization[31] and on the stereochemistry of polymerization[32] has been extensively studied. The effect of $G^{\ominus}$ on the molecular weights of PIB at different temperatures has been reported by Imanishi[12, 18] and Kennedy and co-workers[4, 10, 17]. Imanishi[12, 18] used "H_2O"/$TiCl_4$, CCl_3COOH/$TiCl_4$ and CCl_3COOH/$SnCl_4$ initiator system, n-C_6H_{14}, $CHCl_3$ and CH_2Cl_2 solvents at -20 °C, -50 °C and -70 °C. While molecular weight comparison has not been carried out, k_{tr}/k_p was reported to be greater than k_t/k_p, and larger for $SnCl_4$ than for $TiCl_4$ systems. Increase in solvent dielectric constant increased k_{tr}/k_p, while k_t/k_p was less affected. These results were explained on the basis of a transfer mechanism, similar to that proposed by Kennedy and Thomas[13]. Free ions were thought to be more important for transfer to monomer than propagation. Thus increase in $G^{\ominus}$ size ($SnCl_4 \cdot CCl_3COO^{\ominus} > TiCl_4 \cdot CCl_3COO^{\ominus}$) or solvent polarity increased free ion contribution leading to an increase in k_{tr}/k_p.

Kennedy, Shinkawa and Williams[10] polymerized isobutylene by γ-ray radiation in bulk and compared $\bar{M}_n$, $\bar{M}_v$ and $\bar{M}_w$ of the product with $\bar{M}_v$ of PIB prepared with "H_2O"/$AlCl_3$, BF_3, and $EtAlCl_2$/MeCl. Molecular weights decreased as: γ-rays $>$ $EtAlCl_2 > BF_3 > AlCl_3$ and $\Delta E_{\bar{M}_v} = -6.6$ kcal/mole suggesting the presence of similar molecular weight governing mechanisms. The authors concluded that since highest molecular weights were obtained by γ-radiation, i.e., in absence of counteranion, the least encumbered ion pair would yield the highest molecular weight. The authors also concluded that the presence of a counteranion impedes propagation to a greater extent than transfer and to achieve higher molecular weights with chemical initiators, a weakly nucleophilic counteranion should be used. Kennedy et al.[10], thus arrived at an exactly opposite conclusion than Imanishi[12, 18] in regard to the effect of counteranion on PIB molecular weight.

Kennedy and Rengachary[4] found that PIB molecular weight decreased as: $Me_2AlCl_2^\ominus > Me_2AlClBr^\ominus > Et_2AlCl_2^\ominus > Et_2AlClBr^\ominus > MeAlCl_2OH^\ominus > Et_3AlBr^\ominus > Et_3AlCl^\ominus$ (Fig. 17).

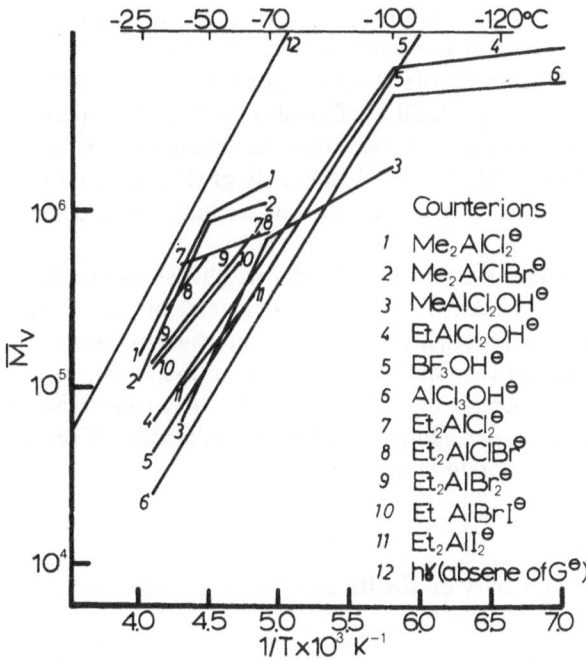

Fig. 17. Effect of temperature on $\overline{M}_v$ in isobutylene polymerization

A combination of all these data[4, 10, 17] is shown in Fig. 17. For the systems with $\Delta E_{\overline{M}_v} = -6.6 \pm 1$ kcal/mole (transfer dominated), $\overline{M}_v$'s decrease depending on $G^\ominus$ as follows: γ-radiation (absence of $G^\ominus$) $> Me_2AlCl_2^\ominus > Me_2AlClBr^\ominus > MeAlCl_2OH^\ominus > EtAlCl_2OH^\ominus > BF_3OH^\ominus > AlCl_3OH^\ominus$. Clearly, highest molecular weights were obtained in the absence of $G^\ominus$, however, the above order does not follow a mere change in the size of $G^\ominus$.

As reported earlier $\overline{M}_v$ of PIB prepared using t-BuX/Et$_2$AlX/MeX, was found to decrease as: $Et_2AlCl_2^\ominus > Et_2AlClBr^\ominus > Et_2AlBr_2^\ominus > Et_2AlClI^\ominus \cong Et_2AlBrI^\ominus > Et_2AlI_2^\ominus$. In this series of counteranions, the number of alkyl groups is constant and the size of $G^\ominus$ is affected only by a change in halogen. These results prove that the $\overline{M}_v$ decreases as the size of $G^\ominus$ increases and do not follow predictions of Kennedy and co-workers[10] according to which $\overline{M}_v$ should increase with the increasing $G^\ominus$ size.

The above order with $Et_2AlX_2^\ominus$ counteranions as well as that developed on the basis of data in Fig. 17 can be understood by the following considerations. The molecular weight of PIB is determined by the nucleophilicity of counteranion and an increase in nucleophilicity tends to decrease the rate of transfer to monomer and to increase the rate of termination. The nucleophilicity of counteranion in turn is determined by the extent of charge delocalization, which is mainly governed by the number and nature of halogens and/or strongly electronegative groups like $-OH$, and much less by alkyl groups. The overall size of $G^\ominus$ is important only insofar as it affects the extent of charge delocalization. For counteranions with an equal number of similar alkyl groups, nucleophilicity is determined solely by the nature of the halogens. Thus, with counteranions with equal number of halogens, the nucleophilicity decreases leading to a decrease in molecular weights in the order: $Me_2AlCl_2^\ominus > Me_2AlClBr^\ominus$; $BF_3OH^\ominus > AlCl_3OH^\ominus$ ($-OH$ is introduced by "H_2O"); $Et_2AlCl_2^\ominus > Et_2AlClBr^\ominus > Et_2AlBr_2^\ominus > Et_2AlClI^\ominus \cong Et_2AlBrI^\ominus > Et_2AlI_2^\ominus$. These sequences strongly indicate that an increase (decrease) in nucleophilicity of $G^\ominus$ increases (decreases) PIB molecular weights. The nucleophilicity of $G^\ominus$ is determined not by the actual but by its "effective" size; i.e., the area over which charge delocalization occurs.

It follows then that for the systems in which molecular weights are controlled by transfer to monomer ($\Delta E_{\overline{M}_v} = -6.6 \pm 1.0$ kcal/mole), an increase in counteranion nucleophilicity increases PIB molecular weights. For systems in which molecular weight control is by a combination of termination and transfer to monomer, ($\Delta E_{\overline{M}_v} = -4.6 \pm 1.0$ kcal/mole) an increase in nucleophilicity leads to an increase in PIB molecular weight, e.g. $Et_2AlX_2^\ominus$ series, indicating that transfer is more affected than termination by a change in counteranion nucleophilicity.

2. Effect of Counteranion Nucleophilicity on the Rate of Transfer in Isobutylene Polymerization

For the system in which transfer to monomer is molecular weight controlling an increase in nucleophilicity of counterions increases the PIB molecular weights. This is clearly shown by the decrease in $\overline{M}_v$ as: γ-irradiation (absence of $G^\ominus$) > $Me_2AlCl_2^\ominus > Me_2AlClBr^\ominus > MeAlCl_2OH^\ominus > EtAlCl_2OH^\ominus > BF_3OH^\ominus > AlCl_3OH^\ominus$. The molecular weight controlling event for all these systems is transfer, as suggested by $\Delta E_{\overline{M}_v} = -6.6$ kcal/mole.

Transfer to monomers is proposed to take place by a concerted mechanism with counteranion assistance[13]:

Kennedy et al.[10] as well as Hayes and Pepper[33] have pointed out that transfer is entropy driven while propagation, with an unfavorable entropy change, is enthalpy driven process. In fact, unavoidable steric compression in the transition state affects transfer more than propagation[33]. These findings support the assumption that a more nucleophilic counteranion decreases the rate of transfer because it results in a more compressed and hence less favorable transition state, thereby leading to an increase in PIB molecular weights. It is speculated that an increase in counteranion nucleophilicity brings about an unfavorable entropy change for the transfer mechanism leading to reduced rate of transfer.

In line with Higashimura's view[34] that carbenium ions are not strictly sp^2 hybridized and that they retain some sp^3 character, a more nucleophilic $G^{\ominus}$ would be expected to induce more sp^3 character to the growing cation. Pronounced sp^3 character of the carbenium ion would prevent orbital overlap, i.e., the formation of transition state leading to transfer, and thus to increase in molecular weight.

In the absence of counteranions, as in irradiation induced polymerization, the transition state for transfer is least favorable, which leads to highest molecular weights. Kennedy[35] has proposed a possibility of four membered transition state of transfer, which can take place in the absence of counteranion assistance.

3. Effect of Counteranion Nucleophilicity on the Rate of Termination in Isobutylene Polymerization

Though increased $G^{\ominus}$ nucleophilicity tends to decrease the rate of transfer, it is also expected to increase the rate of termination. For the systems characterized by $\Delta E_{\overline{M}_v} = -1.8 \pm 1.0$ kcal/mole, however, transfer is negligible and termination controls molecular weights.

Kennedy and Milliman[3] have shown that for systems with $\Delta E_{\overline{M}_v} = -1.8 \pm 1.0$ kcal/mole, $\overline{M}_v$ decreases as $Et_2AlCl_2^{\ominus} > Me_3AlCl^{\ominus} > Et_3AlCl^{\ominus} > i\text{-}Bu_3AlCl^{\ominus}$. Termination is by hydridation by $Et_2AlCl_2^{\ominus}$, $Et_3AlCl^{\ominus}$ and $i\text{-}Bu_3AlCl^{\ominus}$ and methylation by $Me_3AlCl^{\ominus}$ [4]. Among these four $G^{\ominus}$, $Et_2AlCl_2^{\ominus}$ is the "least terminative" conceivably because it is least nucleophilic and hence yields highest molecular weights. Termination is slow also with $Me_3AlCl^{\ominus}$ because it can occur only by a relatively slow $CH_3^{\ominus}$ migration (methylation), whereas with $Et_3AlCl^{\ominus}$ and $i\text{-}Bu_3AlCl^{\ominus}$ termination is by rapid $H^{\ominus}$ transfer (hydridation). That hydridation is much faster than alkylation has been shown by Kennedy and Rengachary[4].

A comparison of $\overline{M}_v$'s obtained with $Me_3AlCl^{\ominus}$ and $Me_2AlCl_2^{\ominus}$ as well as $Et_3AlCl^{\ominus}$ and $Et_2AlCl_2^{\ominus}$ is of interest (Fig. 17). In comparison to $Me_3AlCl^{\ominus}$, the higher stability of $Me_2AlCl_2^{\ominus}$ due to greater charge delocalization reduces the nucleophilicity to such a degree that termination is suppressed and chain transfer becomes molecular weight controlling ($\Delta E_{\overline{M}_v} = -6.6 \pm 1.0$ kcal/mole). The higher $\overline{M}_v$'s obtained with $Et_2AlCl_2^{\ominus}$ than with $Et_3AlCl^{\ominus}$ are simply due to a decreased rate of termination because of decreased nucleophilicity. While $Et_2AlCl_2^{\ominus}$ is more apt to terminate (by hydridation), $Me_2AlCl_2^{\ominus}$ rather leads to transfer, even though the extent of charge delocalization is similar in both $G^{\ominus}$'s. Conceivable, rapid termination by hydridation with the Et-containing counteranion causes this difference.

Thus, to obtain high PIB molecular weight with chemical initiator systems, highly nucleophilic and highly stable counteranions should be used. Counteranions such as $Me_2AlClF^\ominus$ or $Me_2AlF_2^\ominus$ may go a long way toward "living" cationic polymerization since they would be more nucleophilic than $Me_2AlCl_2^\ominus$ and more stable towards termination than $Et_2AlCl_2^\ominus$.

4. Effect of Solvent and Temperature on Molecular Weights in Isobutylene Polymerizations

Since change in nucleophilicity of $G^\ominus$ leads to a change in PIB molecular weight, for a given $G^\ominus$ a change in the ionic interaction should also bring about molecular weight changes. Thus, it is conceivable that by increasing solvation *i.e.*, reducing cation-anion proximity and hence termination by $G^\ominus$ a counteranion can be rendered more transfer-active. In contrast, a less solvating medium should decrease ion separation and thereby decrease transfer.

Such solvent effect are indeed found. As reported above for the t-BuCl/Et_2AlCl/ MeCl system, $\Delta E_{\bar{M}_v} = -1.8 \pm 1.0$ kcal/mole while for the t-BuCl/Et_2AlCl/MeBr, $\Delta E_{\bar{M}_v} = -4.6$ kcal/mole in the range from $-30°$ to $-45 °C$. Thus a change of solvent from MeCl to MeBr increased $\Delta E_{\bar{M}_v}$ and induced transfer, though the $G^\ominus$ in both cases was $Et_2AlCl_2^\ominus$. Since MeBr is more polarizable and of higher donicity than MeCl, it is a better solvating agent of carbenium ions.

By decreasing solvation ionic interaction increases and termination is induced. This would show up in a decrease in $\Delta E_{\bar{M}_v}$. Thus, while Kennedy and co-workers[10, 17] have shown that $\Delta E_{\bar{M}_v} = -6.6 \pm 1.0$ kcal/mole (*i.e.*, transfer active) for "H_2O"/$AlCl_3$, BF_3 or $EtAlCl_2$/MeCl systems, $\Delta E_{\bar{M}_v} = -4.6 \pm 1$ kcal/mole for "H_2O"/$AlCl_3$/ propane[25]; "H_2O"/$EtAlCl_2$/n-hexane[11], "H_2O"/$EtAlCl_2$/n-pentane (this work) and "H_2O"/BF_3/bulk monomer[37, 38] (Table 8). The decrease of $\Delta E_{\bar{M}_v}$ from -6.6 ± 1.0 kcal/mole to -4.6 ± 1 kcal/mole indicates that molecular weight control changes from transfer to monomer to a combination of transfer and termination, which in turn indicates an increase in ionic interactions.

The effect of temperature on $\bar{M}_v$ has been studied by a number of workers (Table 8) and in all cases, a decrease in temperature increased PIB molecular weight. Since solvent dielectric constant increases with decreasing temperatures, molecular weights also are expected to decrease. Apparently such effect is small as shown by the increase in $\bar{M}_v$'s with decreasing temperature. At very low temperatures, $\bar{M}_v$ suddenly drops as shown above. This was explained[4] by assuming a reduced rate of initiation leading to an increase in transfer to initiator.

In sum, a relation of counteranion nucleophilicity and the molecular weight in isobutylene polymerization is discovered, according to which an increase in $G^\ominus$ nucleophility leads to an increase in the rate of termination but a decrease in the rate of chain transfer to monomer. Thus, an increase in $G^\ominus$ nucleophilicity leads to increased termination and hence decreased molecular weight for systems in which termination is molecular weight governing. Similarly, it leads to a decrease in rate of transfer and hence to an increase in molecular weights for systems in which chain transfer controls molecular weight. The nucleophilicity of $G^\ominus$ is determined by the extent of charge delocalization, which in turn is governed by strongly electronegative groups like halogens and −OH. The effect of solvent on molecular weights or $\Delta E_{\overline{M}_v}$ can also be accounted for by a change in the ionic interaction due to a change in solvation. These effects of $G^{\ominus}$ and solvent on molecular weights may be applicable to other cationic olefin polymerization systems in which transfer and termination mechanisms are similar to those operating in isobutylene polymerization.

VII. Conclusions

The effect of t-BuX, Et_2AlX or $EtAlCl_2$, MeX and temperature on PIB molecular weight and molecular weight distribution has been studied by GPC and viscometry. A large amount of data on $\overline{M}_n$, $\overline{M}_w$, MWD as well as $\overline{M}_v$ and $\Delta E_{\overline{M}_v}$ have been generated.

GPC analysis of PIB's prepared using t-BuX/Et_2AlX/MeX has shown that while bimodal MWD is obtained with Et_2AlCl, monomodal MWD's are found with Et_2AlBr and Et_2AlI. The effect of temperature and monomer concentrations on $\overline{M}_n$, $\overline{M}_w$ and MWD of total polymer, HMWF and LMWF has been determined. Mayo plots using $\overline{M}_n$'s of HMWF of PIB prepared with t-BuCl or t-BuBr/Et_2AlCl/MeCl at −50° or −60 °C have greatly increased our understanding of molecular weight controlling mechanisms. For the above systems, it has been found that termination is molecular weight controlling while transfer to monomer is negligible. It is postulated that bimodal MWD's obtained with Et_2AlCl coinitiator system are due to the presence of an impurity, probably water.

Similarly, the effect of t-BuX, Et_2AlX or $EtAlCl_2$/MeX and temperature on $\overline{M}_v$ of PIB has been studied in great detail. In general, decreasing temperatures increased $\overline{M}_v$. Based on Arrhenius plots, a large number of $\Delta E_{\overline{M}_v}$ values have been determined for the above systems. These, together with those reported in the literature are given in Table VIII. The effect of monomer concentration on $\overline{M}_v$ of PIB prepared with t-BuCl or t-BuBr/Et_2AlCl/MeCl systems in the range from −30° to −60 °C has been determined and relative rate constants, k_t/k_p and k_{tr}/k_p, have been calculated from Mayo plots. For t-BuCl/Et_2AlCl/MeCl, termination is the predominant molecular weight governing mechanism while for t-BuBr/Et_2AlCl/MeCl a combination of transfer to monomer and termination is molecular weight controlling, though termination is found to be more important at lower temperature, $i.e.$, −50 °C.

A thorough examination of $\Delta E_{\overline{M}_v}$ data has shown that the $\Delta E_{\overline{M}_v}$ values fall into three groups, characterized by (kcal/mole), −6.6 ± 1.0, −4.6 ± 1.0 and −1.8 ± 1.0. A hypothesis suggesting a relation between the $\Delta E_{\overline{M}_v}$ and molecular weight governing mechanisms has been proposed, which state that $\Delta E_{\overline{M}_v}$'s = −6.6 ± 1.0, −4.6 ± 1.0 and −1.8 ± 1.0 reflect three different molecular weight governing mechanisms: transfer to monomer, a combination of transfer to monomer and termination, and termination, respectively. Experimental proofs in support of this hypothesis have been obtained by Mayo plot analyses and model experiments.

Based on the relation between $\Delta E_{\overline{M}_v}$'s and molecular weight governing events and the $\overline{M}_v$ data obtained in this study as well as those reported in the literature, the effect of counteranions and solvents on molecular weight in isobutylene polymerization has been explained.

Acknowledgements. Financial help by the National Science Foundation and the Firestone Tire and Rubber Company is gratefully acknowledged.

VIII. References

1) Kennedy, J. P., Trivedi, P. D.: Adv. Polymer Sci. *28*, 1 (1978)
2) Kennedy, J. P., Trivedi, P. D.: Polymer Preprints *17*, (2), 791 (1976)
3) Kennedy, J. P., Desai, N. V., Sivaram, S.: J. Am. Chem. Soc. *95*, 6386 (1973)
4) Kennedy, J. P., Rengachary, S.: Adv. Polymer Sci. *14*, 1 (1974)
5) Morton, M., Fetters, L. J.: Rubber Chem. and Tech. *48*, (3), 359 (1975)
6) Masuda, T., Higashimura, T.: J. Polymer Sci. *B, 9,* 793 (1971)
7) Higashimura, T., Kishiro, O.: J. Polymer Sci. *A-1, 12,* 967 (1974)
8) Irie, M., Hayashi, K.: Progress in Polymer Sci. Japan *8,* 105 (1975)
9) Higashimura, T., Kishiro, O.: Polymer J. *9,* (1), 87 (1977)
10) Kennedy, J. P., Shinkawa, A., Williams, F.: J. Polymer Sci. *A-1, 9,* 1551 (1971)
11) Maslinska-Solich, J., Chmelir, M., Marek, M.: Collection Czechoslov. Chem. Commun. *34,* 2611 (1969)
12) Imanishi, Y., Higashimura, T., Okamura, S.: Kobunshi Kagaku *18,* 333 (1961)
13) Kennedy, J. P., Thomas, R. M.: J. Polymer Sci. *49,* 189 (1961)
14) Flory, P. J.: J. Am. Chem. Soc. *65,* 372 (1943)
15) Flory, P. J.: Principles of polymer chemistry. New York: Cornell Univ. Press 1953
16) Biddulph, R. H., Plesch, P. H., Rutherford, P. P.: J. Chem. Soc. *1965,* 275
17) Kennedy, J. P., Squires, R. G.: Polymer *6,* 579 (1965)
18) Imanishi, Y.: Ph. D. Thesis, Kyoto Univ. 1964
19) Plesch, P. H.: The chemistry of cationic polymerization. New York: Pergamon Press Book 1963
20) Cesca, S., Giusti, P., Magagnini, P., Priola, A.: Makromol. Chem. *176,* 2319 (1975)
21) Kennedy, J. P., Trivedi, P. D.: XXIII IUPAC Meeting on Macromolecules, Madrid, *1,* 198 (1974)
22) Mayo, F. R., Gregg, R. A., Matheson, M. S.: J. Amer. Chem. Soc. *73,* 1691 (1951)
23) Pác, J., Plesch, P. H.: Polymer *8,* 237 (1967)
24) Maina, M., Cesca, S., Giusti, P., Ferraris, G., Magagnini, P. L.: Makromol. Chem. *178,* 2223 (1977)
25) Kennedy, J. P., Thomas, R. M.: Adv. Chem. Ser. *34,* 111 (1962)
26) Pepper, D. C.: Quart. Rev. *8,* 90 (1954)
27) Sawada, H.: J. Macromol. Sci.-Revs. *C7,* 161 (1972)
28) Kennedy, J. P., Huang, S. Y., Feinberg, S. C.: J. Polymer Sci. Chem. *15,* 2801 (1977)
29) Toman, L., Marek, M.: Makromol. Chem. *177,* 3325 (1976)
30) Priola, A., Cesca, S., Ferraris, G., deMaina, M.: Makromol. Chem. *176,* 2289 (1975)
31) Kennedy, J. P., Johnston, J. E.: Adv. Polymer Sci. *19,* 57 (1975)
32) Kunitake, T.: Ionic polymerization – Unsolved problems. Furukawa, J., and Vogl, O. (Ed.) Marcel Dekker, Inc. 1976
33) Hayes, M. J., Pepper, D. C.: Proc. Roy. Soc. (London) *A 263,* 63 (1961)
34) Higashimura, T., Yonezawa, T., Okamura, S., Fukui, K.: J. Polymer Sci. *49,* 487 (1959).
35) Kennedy, J. P.: XXIVth IUPAC Symposium, Hamburg, W. Germany, *1,* 25 (1974)
36) Kennedy, J. P., Milliman, G. E.: Advances in Chemistry Series *91,* 287 (1969)
37) Thomas, R. M., Sparks, W. J., Frolich, P. K., Otto, M., Mueller-Cunradi, M.: J. Amer. Chem. Soc. *62,* 276 (1940)

[38] Kennedy, J. P., XXIII Inter. Congress of Pure Applied Chem.: Macromolecular Preprint *1*, 105 (1971)

[40] Addecott, K. S. B., Mayor, L., Turner, C. N.: European Polymer J. *3*, 601 (1967)

[41] Cesca, S., Priola, A., Bruzzone, M., Ferraris, G., Giusti, P.: Makromol. Chem. *176*, 2339 (1975)

[42] Chmelir, M., Marek, M., Wichterle, O.: J. Polymer Sci. *C 16*, 833 (1967)

[43] Kennedy, J. P., Sivaram, S.: J. Macromol. Sci.-Chem. *A 7* 969 (1973)

[44] Giusti, P., Priola, A., Magagnini, P., Narducci, P.: Makromol. Chem. *176*, 2303 (1975)

[45] Mandal, B. M., Kennedy, J. P.: J. Polymer Sci.-Chem. *16*, 833 (1978)

Received January 18, 1978
H.-J. Cantow (editor)

Author Index Volumes 1–28

Polymers

Properties
and
Applications

Volume 1

B. Rånby, J.F. Rabek

ESR Spectroscopy in Polymer Research

1977. 356 figures, 29 tables. XIV, 410 pages
ISBN 3-540-08151-8

The main purpose of this book is to collect the present available information on the applications of electron spin resonance (ESR) spectroscopy in polymer research. The book has been written both for those who want an introduction to this field, and for those who are already familiar with ESR and are interested in application to polymers. Therefore, the fundamental principles of ESR spectroscopy are first outlined, the experimental methods including computer applications are described in more detail, and the main emphasis is on the application of ESR methods to polymer problems. The authors hope that this book will provide a useful source of information by giving a coherent treatment and extensive references to original papers, reviews, and discussions in monographs and books. In this way we hope to encourage polymer chemists, organic chemists, biochemists, physicists, and material scientists to apply ESR methods to their research problems. (2519 references).

Volume 2

H.-H. Kausch

Polymer Fracture

1978. Approx. 350 pages
ISBN 3-540-08786-9

In the last fifteen years modern spectroscopical methods (ESR, IR) and conventional methods of structure research have permitted considerable progress in the investigation of deformation and fracture of polymeric materials. For the first time in western languages a unified view of the kinetic theory of polymer fracture is presented by one of the scientists contributing to its development.

Springer-Verlag
Berlin
Heidelberg
New York

Polymer Bulletin

Editors:

Prof. H.-J. Cantow
Institute of Macromolecular
Chemistry
University of Freiburg
Stefan-Meier-Strasse 31
D-78 Freiburg/Germany

Prof. J.P. Kennedy
Dept. of Polymer Science
The University of Akron
Akron, OH 44325/USA
Prof. T. Saegusa
Dept. of Synthetic Chemistry
Kyoto University
Kyoto, 606 Japan

The articles are to be sent to one of the editors or to
Springer-Verlag Berlin Heidelberg New York

Polymer Bulletin

Preface

To cope with the rapid progress of polymer science, a new
journal is now published characterized by emphasis on rapid
publication of papers containing a most concise description of
results.
The character of the new journal is between the purely archival
journal of full papers and the so-called "letter journals" consisting exclusively of short communications.

Ask for our detailed leaflet!

The journal consists of one volume a year, published in 12
issues.
Subscription information upon request.

Springer
International